ΛUTO RICERCA

Subtle energies and vibrational states

Issue 20
Year 2020

Journal	AutoRicerca
Issue	20
Year	2020
Pages	296
ISSN	2673-5105
Title	Subtle energies and vibrational states
Authors	Alegretti, W., Montenegro, R., Sassoli de Bianchi, M. & Trivellato, N.
Editor	Massimiliano Sassoli de Bianchi
Cover	Luca Sassoli de Bianchi
Copyright	The authors (all rights reserved)
Web	www.autoricerca.ch, www.autoricerca.com

AutoRicerca is a publication of the *LAB – Laboratorio di AutoRicerca di Base* (www.autoricerca.ch), c/o *Area 302 SA* (www.area302.ch), via Cadepiano 18, 6917 Barbengo, Switzerland.

Index

The pages of a book, whether paper or electronic, possess a peculiar property: they are able to accept whatever variety of letters, words, phrases and illustrations, without ever expressing a criticism, or disapproval. It is important to be aware of this fact when we go through a text, so that the lantern of our discernment can always accompany our reading. To explore new possibilities, we must remain open-minded, but it is equally important not to succumb to the temptation to uncritically absorb everything we read. In other words, the warning is to always subject the content of our reading to the scrutiny of our critical sense and personal experience. The editor and the authors can in no way be held responsible for the consequences of a possible paradigm shift induced by the reading of the words contained in this volume.

Le pagine di un libro, siano esse cartacee o elettroniche, possiedono una particolarissima proprietà: sono in grado di accettare ogni varietà di lettere, parole, frasi e illustrazioni, senza mai esprimere una critica, o una disapprovazione. È importante essere pienamente consapevoli di questo fatto, quando percorriamo uno scritto, affinché la lanterna del nostro discernimento possa accompagnare sempre la nostra lettura. Per esplorare nuove possibilità è indubbiamente necessario rimanere aperti mentalmente, ma è ugualmente importante non cedere alla tentazione di assorbire acriticamente tutto quanto ci viene presentato. In altre parole, l'avvertimento è di sottoporre sempre il contenuto delle nostre letture al vaglio del nostro senso critico ed esperienza personale. L'editore e gli autori non possono in alcun modo essere ritenuti responsabili circa le conseguenze di un cambiamento di paradigma indotto dalla lettura delle parole contenute in questo volume.

Editorial

[*Per la versione in italiano, vai a pagina 10*]

After 9 years of life, more than 4,700 pages, 25 articles, 11 monographic texts and 18 authors (not counting the anonymous ones), with this new issue, the first of 2020, *AutoRicerca* changes its look, both with regards to its covers and the internal layout. A change that marks the journal's intention to continue to grow in the years to come, by (1) addressing an increasingly international audience and (2) proposing more frequently also themes not necessarily specifically related to inner research.

Concerning point (1), I draw attention to the fact that of the 19 volumes published from 2011 to 2019, four have already been published also as an English edition (more precisely, volumes 7, 12, 16 and 19), whereas one was published exclusively in English (volume 18). This is to underline that *AutoRicerca*, although born as an Italian journal, has always tried, within the limits of its resources, to communicate with the widest possible readership. In the years to come, the intention is to continue to offer contents also in English with greater frequency, in parallel to the Italian editions, or as stand-alone volumes only in English.

This is the case of the present volume, which closes the cycle of the first 20 issues of the journal, proposing in English some of the articles that were previously published only in Italian, plus some extra content. More precisely, in this volume you will find the English version of 6 articles that were published in volumes 1, 6 and 10, by *Wagner Alegretti*, *Nanci Trivellato* and *Massimiliano Sassoli de Bianchi*, plus two additional articles, one again by Trivellato, with also a review of a recent book published by this author, and the other one written by *Rodrigo Montenegro*.

The entire volume revolves around the theme of so-called "subtle energies," and more precisely the problem of their objective detection, their correct conceptualization, as well as the possibility of "setting them into motion" in the context of specific inner techniques, aimed at obtaining a condition called the *vibrational state* (VS). I hope that the richness and interest of these contents will make this volume a future reference on the delicate and controversial topic of "bioenergy."

As regards the aforementioned point (2), *AutoRicerca* already dealt in the past with topics beyond inner research per se, being clear that any inner knowledge requires articulation and integration – possibly coherent and harmonious – with knowledge coming from the observation and study of the external reality as well. In this regard, volumes 2, 18 and 19 of the journal already dealt with the theme of physics, while volume 13 addressed the controversial subject of disease and its possible causes (although from an undoubtedly unconventional perspective).

This openness of *AutoRicerca* towards contents that are not exclusively about inner/spiritual research will from now on be clearly expressed also in the new covers, with the previous wording "Journal of inner research" replaced by the more general wording: "Journal of inner & outer research." In this regard, I can already anticipate that the next Issue 21, which is in preparation, will be entirely devoted to the theme of physics, in particular quantum physics and relativity, demonstrating once more this "all-round" view of the journal on the world of research and dissemination of knowledge.

The restyling of the journal also anticipates a new phase of the *Laboratorio di Autoricerca di Base* (LAB), which towards the end of 2020 will have again a stable physical location, in the premises of *Area 302 SA*, in Lugano (Switzerland), where it will be possible to organize again conferences, courses, seminars and moments of practice. To find out more, stay tuned via the LAB's website (*www.autoricerca.ch*) and/or by subscribing to its mailing list.

As the journal of the LAB, the mission of *AutoRicerca* will remain that of furthering competences and knowledge that can maximize human potential. I would add that the volumes of the journal will remain in the future of a thematic nature and that the new graphic of its covers will put more emphasis on this aspect, giving more importance to the title of the volume with respect to the name of the journal. Furthermore, the authors' names will from now on also appear on the cover, to give more value to their precious contribution.

But let me come now to the content of this issue. For the first three articles, let me repeat here what I wrote (in Italian) in the editorial of issue Number 1, back in 2011.

The first article, by Wagner Alegretti, examines the nature of the *vibrational state* (VS) from the viewpoint of its possible effects on the neurophysiology of the physical brain. The author presents in his work a particular study protocol aimed at identifying and analyzing the possible *neural correlates* of the VS. This not only to bring this particular energy condition to a higher degree of objectivity, so that it can also be studied by conventional neuroscientists (like any other state of consciousness, such as lucid dreams and meditative states), but also and above all to deepen our knowledge about this particular multidimensional phenomenon, improving the future paratechnologies (i.e., inner technologies) able to produce and control it. The first part of Alegretti's article also contains an introduction to the VS, as well as an explanation of the technique of the *voluntary energetic longitudinal oscillation* (VELO), which is able to promote it. Alegretti's article will therefore be useful also for those readers who are approaching this phenomenon for the first time.

The second article, by Nanci Trivellato, presents a detailed analysis of the attributes underlying the aforementioned VELO technique. The study by Trivellato identifies and analyzes in detail four different categories of attributes, which the practitioner can advantageously learn to recognize and master, in order to promote (through the firm application of her/his will) vibrational states of ever greater intensity and depth. More precisely, the author identifies in her analysis 5 primary energetic attributes, 5 derived energetic attributes, 8 composite energetic attributes and 3 intercurrent intraconsciential attributes, for a total of 21 fundamental identifiable attributes.

Finally, the third article, by Massimiliano Sassoli de Bianchi, illustrates some interesting parallels between the so-called consciential paradigm, as it was defined in the approach known as *Conscientiology*, and the metaphysical vision which is at the basis of *Yoga*. As pointed out in the article, advanced consciousnesses have probably walked this planet in ancient times, leaving as a legacy knowledge of considerable value, such as the ancient breathing (*pranayama*) and energetic (*pranavidya*) techniques present in the body of teachings of Yoga. Regarding these techniques, the text proposes to integrate a particular yogic respiratory procedure, called *circular breathing* (little known also in the ambit of Yoga) with the aforementioned VELO procedure, with which it shares some interesting structural similarities. This in order to obtain a hybrid procedure, or technique, of a physico-energetic nature, which allows for a gradual approach to the development of an energosomatic mastery.

The content of the next article, by Wagner Alegretti, was presented by the author during the *International Congress of Conscientiology – consciousness science –* held on 22-24 May 2015, in Portugal, at the Campus of the *International Academy of Consciousness* (IAC). The author describes in the article some preliminary results on the detection of "subtle bioenergetic fluxes" exteriorized by Alegretti himself, using *functional magnetic resonance imaging* (fMRI). Undoubtedly, if the unexpected results he obtained will received further confirmation, these could change the current "rules of the game." In fact, what is still lacking today in the investigation of the "subtle energy substances" is the possibility of detecting them directly and in a replicable way, using specific instruments. Of course, this possibility is not only of interest for inner research, as a confirmation of the authenticity of certain paraperceptions (subtle perceptions), but also and above all for outer research, especially for physics. In fact, the objective experimental discovery of new fields of matter-energy (in this case, bridging the physical and the extra-physical domains) immediately raises the problem of identifying their nature, how they could be modeled, their relationship/interaction with other known physical fields, etc.

An additional article by Nanci Trivellato follows, the content of which was also presented during the aforementioned congress, in which the author clarifies numerous misunderstandings and confusions about the VS phenomenon, presenting results obtained in many years of personal research, resulting also from her analysis and accurate evaluation of the energetic sphere of nearly a thousand practitioners of the technique, allowing Trivellato to propose specific measurement scales for the phenomenon.

Then, we have a very interesting research by Rodrigo Montenegro, on the difficult problem of identifying the neurophysiological correlates of the VS and its accompanying states, such as the Out-of-Body Experience (OBE). More precisely, in this preliminary report, Montenegro elaborates a critical framework to technically define the VS (usually only understood as a non-ordinary experiential state of consciousness) also as a characteristic and unmistakable neurophysiological state.

The article will not be an easy read for the general reader without a smattering of the neurology and anatomy-physiology of the brain, but one can also follow the different reasonings of the author regardless of the specific details, and in this way appreciate the complexity of the problem, when it comes to identifying new states of consciousness and, more importantly, properly distinguish them from the other known states.

The two concluding articles, by Massimiliano Sassoli de Bianchi, were previously published in Italian in Volume 6 of *AutoRicerca*. These are texts that address the theme of *energy* mostly from a theoretical point of view. In the first, the author offers an important conceptual clarification about the very notion of energy, exposing numerous commonplaces, often spread by the same experts. For example: is it really true that there are different forms of energy?

The last article addresses the question of the known gap between the physical and extraphysical dimensions. Here the author develops an ultra-simplified model – a sort of scientific metaphor – capable if not of explaining at least of illustrating the reasons why an extraphysical substance would usually interact so weakly (except in special circumstances) with a physical substance, and vice versa.

At the end of this rich in content volume, the reader will also find a review of an interesting and important book published in 2017 by Nanci Trivellato, which embodies in a systematic and precise exposition the knowledge accumulated so far by the author on the techniques that are able to favor the VS.

As always, the wish is for a good read and study and, above all, for a good practice!

Massimiliano Sassoli de Bianchi
Editor

Editoriale

Dopo 9 anni di vita, più di 4'700 pagine, 25 articoli, 11 testi monografici e 18 autori (senza contare quelli anonimi), con questo nuovo numero, il primo del 2020, *AutoRicerca* cambia look, sia per quanto riguarda le sue copertine che per il layout interno. Un cambiamento che marca l'intenzione della rivista di continuare a crescere negli anni a venire: (1) rivolgendosi a un pubblico sempre più internazionale e (2) proponendo con sempre maggiore frequenza anche tematiche non necessariamente specificatamente legate alla ricerca interiore.

Riguardo il punto (1), attiro l'attenzione sul fatto che dei 19 numeri pubblicati dal 2011 al 2019, quattro lo sono già stati anche in edizione inglese (più esattamente, i numeri 7, 12, 16 e 19), mentre un volume è stato pubblicato unicamente in lingua inglese (il numero 18). Questo per sottolineare che *AutoRicerca*, pur nascendo come rivista in lingua italiana, ha comunque sempre cercato, nei limiti delle proprie risorse, di comunicare con un pubblico di lettori il più vasto possibile. Negli anni a venire, l'intenzione è di continuare a proporre contenuti anche in lingua inglese, con maggiore frequenza, in parallelo alle edizioni in italiano, o come volumi indipendenti solo in inglese.

È il caso del presente volume, che chiude il ciclo dei primi 20 numeri della rivista, riproponendo in lingua inglese alcuni articoli precedentemente pubblicati solo in italiano, più alcuni contenuti extra. Più esattamente, in questo volume troverete la versione inglese di 6 articoli pubblicati nei numeri 1, 6 e 10, da *Wagner Alegretti*, *Nanci Trivellato* e *Massimiliano Sassoli de Bianchi*, più due articoli aggiuntivi, uno sempre di Trivellato, con anche la recensione di un recente libro scritto da questa autrice, e l'altro scritto da *Rodrigo Montenegro*.

L'intero volume ruota attorno al tema delle cosiddette "energie sottili", e più esattamente il problema della loro rilevazione oggettiva, della loro corretta concettualizzazione, oltre che la possibilità di "metterle in moto" nell'ambito di specifiche tecniche interiori, volte ad ottenere una condizione denominata *stato vibrazionale* (SV). Mi auguro che la ricchezza e interesse di questi contenuti farà di questo volume una futura referenza sul tema delicato e controverso della "bioenergia".

Per quanto riguarda il summenzionato punto (2), già in passato *AutoRicerca* si era occupata di temi al di là della sola ricerca interiore, essendo comunque evidente che ogni conoscenza interiore richieda un'articolazione e integrazione – possibilmente coerente e armonica – con le conoscenze provenienti dall'osservazione e studio anche della realtà esteriore. A questo proposito, già i numeri 2, 18 e 19 della rivista si erano occupati di fisica, mentre il numero 13 ha affrontato il tema controverso della malattia e delle sue possibili cause (sebbene da una prospettiva indubbiamente non convenzionale).

Questa apertura di *AutoRicerca* anche nei confronti di contenuti non esclusivi alla ricerca interiore/spirituale, verrà da ora in poi espressa in modo chiaro anche nelle nuove copertine, con la precedente dicitura "Rivista di ricerca interiore" ora rimpiazzata dalla dicitura più generale: "Rivista di ricerca interiore & esteriore". A riguardo, posso già anticipare che il prossimo Numero 21, in preparazione, sarà interamente dedicato al tema della fisica, in particolar modo la fisica quantistica e la relatività, a dimostrazione ancora una volta dello sguardo "a tutto tondo" della rivista sul mondo della ricerca e della diffusione della conoscenza.

Il restyling della rivista anticipa anche una nuova fase del *Laboratorio di Autoricerca di Base* (LAB), che verso la fine del 2020 disporrà nuovamente di una sede fisica stabile, nei locali di *Area 302 SA*, a Lugano (Svizzera), dove sarà nuovamente possibile organizzare conferenze, corsi, seminari e momenti di pratica. Per saperne di più, rimanete sintonizzati tramite il sito del LAB (*www.autoricerca.ch*) e/o iscrivendovi alla sua mailing list.

In quanto rivista del LAB, la missione di *AutoRicerca* resterà quella della promozione di competenze e conoscenze in grado di valorizzare e massimizzare il potenziale umano. Aggiungo che i volumi della rivista rimarranno in futuro di natura tematica e la nuova grafica delle copertine metterà maggiormente l'accento su questo aspetto, dando più enfasi al titolo del volume rispetto al nome della rivista. Inoltre, i nomi degli autori verranno da ora in poi indicati anche in copertina, per dare più valore al loro prezioso contributo.

Ma veniamo ora al contenuto di questo numero. Per i primi tre articoli, riporto essenzialmente quanto avevo già scritto nell'editoriale del primo numero, nel 2011. Il primo articolo, di Wagner Alegretti, esamina la natura dello stato vibrazionale (SV) dal punto di vista dei suoi possibili effetti sulla neurofisiologia del cervello fisico. L'autore presenta nel suo lavoro un particolare protocollo di studio volto ad individuare e analizzare i possibili *correlati neurali* dello SV. Questo non solo per portare questa particolare condizione energetica a un grado di maggiore oggettività, affinché possa essere studiata anche dai neuroscienziati convenzionali (al pari di altri stati di coscienza, come ad esempio i sogni lucidi e gli stati meditativi), ma anche e soprattutto per approfondire la nostra conoscenza su questo particolare fenomeno multidimensionale, migliorando le future paratecnologie (tecnologie interiori) in grado di produrlo e controllarlo. La prima parte dell'articolo di Alegretti contiene anche un'introduzione allo SV, oltre che una spiegazione della tecnica dell'*oscillazione longitudinale volontaria delle energie* (OLVE), in grado di promuoverlo. La lettura dell'articolo di Alegretti risulterà pertanto utile a quei lettori che approcciano questo fenomeno per la prima volta.

Il secondo articolo, di Nanci Trivellato, presenta un'analisi dettagliata degli attributi alla base della summenzionata tecnica dell'OLVE. Lo studio di Trivellato identifica ed analizza nel dettaglio quattro diverse categorie di attributi, che il praticante potrà vantaggiosamente imparare a riconoscere e padroneggiare, alfine di promuovere (attraverso l'applicazione ferma della propria volontà) stati vibrazionali di sempre maggiore intensità e profondità. Più esattamente, l'autore identifica nella sua analisi 5 attributi energetici primari, 5 attributi energetici derivati, 8 attributi energetici composti e 3 attributi intracoscienziali intercorrenti, per un totale di 21 attributi fondamentali identificabili.

Infine, il terzo articolo, di Massimiliano Sassoli de Bianchi, illustra alcuni interessanti paralleli tra il cosiddetto paradigma coscienziale, così come è stato definito nell'approccio della *Coscienziologia*, e la visione metafisica alla base dello *Yoga*. Come sottolineato nell'articolo, coscienze molto avanzate hanno probabilmente solcato questo pianeta in tempi remoti, lasciando in eredità conoscenze di notevole valore, come ad esempio le antiche tecniche respiratorie (*pranayama*) ed energetiche (*pranavidya*) presenti nel corpus di insegnamenti dello Yoga. Riguardo a queste tecniche, il testo propone di integrare una particolare tecnica respiratoria yogica, detta della *respirazione circolare* (poco conosciuta anche nell'ambito dello Yoga) con la summenzionata tecnica dell'OLVE, con la quale condivide alcune interessanti similarità strutturali. Questo alfine di ottenere una tecnica ibrida, di natura fisico-energetica, che consenta un approccio il più possibile graduale allo sviluppo di una padronanza energosomatica.

Il contenuto dell'articolo successivo, di Wagner Alegretti, è stato presentato dall'autore nel corso dell'*International Congress of Conscientiology – consciousness science –* tenutosi il 22-24 maggio 2015, in Portogallo, presso il Campus dell'*International Academy of Consciousness* (IAC). Nell'articolo l'autore descrive alcuni risultati preliminari sul rilevamento dei "flussi bioenergetici sottili" esteriorizzati dallo stesso Alegretti, tramite *risonanza magnetica funzionale* (fMRI). Indubbiamente, se i risultati inaspettati da lui ottenuti ricevessero ulteriori conferme, potrebbero cambiare le attuali "regole del gioco". Infatti, ciò che oggi ancora manca nell'indagine sulle "sostanze energetiche sottili" è proprio la possibilità di rilevarle in modo diretto e facilmente replicabile, tramite strumentazioni specifiche. Naturalmente, questa possibilità non è di interesse solo per la ricerca interiore, come conferma della veridicità di determinate parapercezioni (percezioni "sottili"), ma anche e soprattutto per la ricerca esteriore, in particolar modo per la fisica. Infatti, la scoperta per via sperimentale oggettiva di nuovi campi di materia-energia (in questo caso, a cavallo tra il fisico e l'extrafisico) pone immediatamente il problema dell'identificazione della loro natura, della loro modellizzazione, della loro relazione/interazione con gli altri campi fisici conosciuti, ecc.

Segue un ulteriore articolo di Nanci Trivellato, i cui contenuti sono stati anch'essi presentati nel corso del summenzionato congresso, in cui l'autrice chiarisce numerosi malintesi e confusioni circa il fenomeno dello SV, presentando i risultati di numerosi anni di ricerca personale, frutto anche di un'analisi e valutazione accurata della sfera energetica di quasi un migliaio di praticanti della tecnica, cosa che le ha permesso di proporre delle specifiche scale di misurazione del fenomeno.

Abbiamo quindi una ricerca molto interessante di Rodrigo Montenegro, sul difficile problema dell'identificazione dei correlati neurofisiologici dello SV e degli stati ad esso associati, come le esperienze extracorporee (OBE). Più precisamente, in questo suo rapporto preliminare, Montenegro elabora un quadro critico per poter definire tecnicamente lo SV (di solito compreso unicamente come stato di coscienza esperienziale non ordinario) anche come stato neurofisiologico specifico e inconfondibile.

L'articolo non sarà di facile lettura per il lettore generico senza un'infarinatura di neurologia e anatomo-fisiologia del cervello, ma si possono sicuramente seguire i diversi ragionamenti dell'autore indipendentemente dai dettagli specifici, e in questo modo apprezzare la complessità del problema quando si tratta di identificare nuovi stati di coscienza e, cosa più importante ancora, distinguerli adeguatamente dagli altri stati noti.

I due articoli conclusivi, di Massimiliano Sassoli de Bianchi, sono stati precedentemente pubblicati in italiano nel Numero 6 di *AutoRicerca*. Si tratta di testi che inquadrano il tema dell'*energia* da un punto di vista soprattutto teorico. Nel primo, l'autore offre un'importante chiarificazione concettuale circa la nozione stessa di energia, smascherando numerosi luoghi comuni, spesso diffusi dagli stessi addetti ai lavori. Ad esempio: è proprio vero che esistono diverse forme di energia?

L'ultimo articolo affronterà invece la questione della nota separazione tra la dimensione fisica e le dimensioni extrafisiche. Qui l'autore svilupperà un modello ultra-semplificato – una sorta di metafora scientifica – in grado se non proprio di spiegare quantomeno di illustrare le ragioni per cui una sostanza extrafisica interagirebbe solitamente così debolmente (salvo circostanze particolari) con una sostanza fisica, e viceversa.

A conclusione di questo ricco volume, il lettore troverà anche la recensione di un interessante e importante libro pubblicato nel 2017 da Nanci Trivellato, che in un'esposizione sistematica e precisa racchiude le conoscenze accumulate sino ad oggi dall'autrice sulle tecniche in grado di favorire la condizione dello SV.

Come sempre, l'augurio è quello di una buona lettura e studio e, soprattutto, di una buona pratica!

Massimiliano Sassoli de Bianchi
Editore

About the authors

Wagner Alegretti, an electronic engineer, has been dedicating to teaching and researching into consciousness since 1980. He has been instrumental in furthering consciousness science. He is a co-founder of the International Academy of Consciousness (IAC), of which he was president from 2001 to 2014, and in establishing IAC's campus in Portugal. He presented at many events, including TEDx and The Science of Consciousness conference (Tucson, AZ, USA). In 1984, Alegretti started working on the development of bioenergy transducers. He continues in his pursuit of finding means for detecting human subtle energy. His recent experiments involve fMRI and other forms of NMR. He authored the book "Retrocognitions," published in English, German, Japanese, Portuguese and Spanish.

Rodrigo Montenegro holds post-graduation diplomas in both Neuropsychology (PgC) and Neuromodulation stimulation (PgC) and has recently studied towards a MSc in Neuroscience under the supervision of Dr Peter Fenwick, a leading expert in the field of Near-Death Experience (NDE). He is currently pursuing a specialization in Sleep Medicine at the University of Oxford, Nuffield Department of Clinical Neurosciences. Montenegro is the author of "Out-of-the-Body Experiences − An Experiential Anthology" and is currently authoring articles in Sleep and circadian rhythm disruptions and neurodegenerative REM sleep behavior disorder synucleinopathy. He has lectured internationally on the topic of consciousness research in-based in neuroscience at the Mind Body and Soul, the Scottish Society of Psychical Research (SSPR), the III Congrès des Thérapies Quantiques, and the Institut de Recherche sur les Expériences Extraordinaires (INREES), among others, since 2005, besides being featured in radio interviews and acclaimed documentaries. He is currently carrying research on the neurophysiological correlates of out-of-body experiences (OBEs). Montenegro is presently the CEO of Gamma Wave Technologies Ltd., a neuroscience-based start-up dedicated to the research of non-ordinary states. Gamma Wave Technologies Ltd. endeavors to apply neuroscience insight to develop transformative technologies in the support real-life needs based on peer-reviewed research and is currently developing a high-quality EEG headband for cognitive enhancement, which among other features will provide neurofeedback data to assess and benchmark the vibrational state of its users, based on current research.

Massimiliano Sassoli de Bianchi received the PhD. degree in physics from the Federal Institute of Technology in Lausanne (EPFL) in 1995, with a study on temporal processes in quantum mechanics. His current research activities are focused on the foundations of physical theories, quantum mechanics, quantum cognition, and consciousness studies (self-research). He is currently a research fellow at the Center Leo Apostel for Interdisciplinary Studies (CLEA), situated at the Vrije Universiteit Brussel (VUB) in Belgium, director of the Laboratorio di Autoricerca di Base (LAB), in Switzerland, editor of its journal AutoRicerca, and president of the company Area 302 SA.

Nanci Trivellato, MSc in Research Methods in Psychology, is a researcher, author, and educator. She is also a charter member of the International Academy of Consciousness (IAC) and the Institute of Applied Consciousness Technologies (I-ACT). She established the Journal of Consciousness, of which she was Chief editor for 15 years, and is the author of the book "Vibrational State and Energy Resonance." She is dedicated to consciousness and multidimensional studies since 1990, and has lectured in Australia, Brazil, Canada, Cyprus, Finland, Germany, Japan, Mexico, Netherlands, Portugal, Romania, Spain, Switzerland, the UK, and the USA. More recently, she presented her research approach in a TEDx Talk entitled "How out-of-body experiences could transform yourself and society", available at: *https://goo.gl/fRQ6c6*.

16

ΛutoRicerca

An approach to the research of the Vibrational State through the study of brain activity

Wagner Alegretti

Issue 20
Year 2020
Pages 17-50

 LΛB

Abstract

This work discusses some possible methodological bases and procedures for the study of the Vibrational State (VS) through direct observation of its neurophysiologic manifestations and energetic repercussions. The author offers a theoretical deepening, allowing for extrapolations, starting from the preliminary studies and hypotheses raised so far. Such research, that allows for replicability of its experiments, is based on the assumption that the VS produces somatic effects, namely in the central nervous system, possible of being detected through adequate technology (i.e., EEG and fMRI). This study aims to address various points not yet answered about the VS, such as: the type of neurological correlates, the brain areas associated with the VS, the identification of how and if the VS differs from other regimes of brain function, the existence (or not) of a pattern of brain waves characteristic of an individual in the VS during the state of coincidence of the vehicles of manifestation, and the mechanisms of the interface between soma and energosoma, among others.

AutoRicerca, Issue 20, Year 2020, Pages 17-50

1 Introduction

Contextualization and history of the research

The author has had interest on the theme here studied for decades, as he has been experiencing out-of-body experiences since childhood, including the vibrational state (VS) and other phenomena related to bioenergy (prana, chi, orgone, vital energy, life force, subtle energy, biofield, and many other expressions), and has always had questions about the mechanism of their occurrence.

For those that have experimented lucid parapsychic and projective experiences, in enough quantity and quality, the reality of the consciousness as being independent from the body, the multidimensionality of the consciousness and the existence of bioenergies are natural facts, meaning, as real as our day to day lives.

Taking into consideration the importance of such knowledge for consciousness study and for the revolutionary and exponential expansion of human knowledge, and knowing of the positive consequences of controlling bioenergies, it is unfortunate that science in general does not yet invest more in the study and furthering of such knowledge. The reasons for such are known and it is not up to this work to discuss the epistemological, political, philosophical, religious, ontological and methodological reasons of the aspects surrounding this question.

It is worth mentioning that there were other studies on some aspects of bioenergies, including some with the purpose of healing. However, the approach and conclusions were in general physicalist and limited, many times hindering more than helping the advancement of multidimensional knowledge and the establishment of a productive interface between this field and other areas of science.

The search for less subjective approaches for the investigation of the multidimensional consciousness and of bioenergies is a noble goal, vital for the development of deeper human knowledge and for the implementation of a new paradigm. Therefore, the fact that conventional science is limited for not having yet been able to comprehend and find ways that allow for a more detailed analysis of the non-physical phenomena point to a condition to be over-come, and there is certainly an opportunity and responsibility to make a contribution in this regard, so that this can happen.

With the goal of establishing a line of research that allowed for the development of parameters and methodology for a bioenergy research, this author presented in 1990, in Rio de Janeiro, during the 1st International Congress of Projectiology, a conference enti-tled "Bioenergetic Technology," where he presented the results of a research done from 1984-1988. That research aimed at identify-ing a relationship that allowed for the establishment of a way for the instrumental detection and measuring of bioenergy (Alegretti, 1990). At the above-mentioned work, the principles of 'bioener-getic technology' were presented, as well the experimental results with a collagen bioenergy transducer, the discussions about its rel-evance and applications and also the planning of the future phases for its development.

In continuation of the study cited above, the bioenergetic re-search of subjective nature (self-research), the field research and the case studies conducted by the author, he developed in 1991 the conceptual and experimental basis of this present work. In that occasion he had the first opportunity of doing some personal experiments with the lucid projection while monitored by an EEG (and other devices for measurement of several physiological signs) in a sleep study laboratory, in the city of Porto Alegre, Brazil. On this occasion, the author also created the Vibrational State volun-tarily, as to allow for the observation of the changes in brain waves pattern that could be generated as a result of this event.

Since then, this author has been actively searching for means and opportunities for the execution of batteries of additional experi-ments where this condition can be examined. So, in December 2007, a series of experimental sessions of EEG analyses were done, having this author and Nanci Trivellato as objects of study, at a neuroscience laboratory in the city of Natal, Rio Grande do Norte, Brazil. Those experiments focused on registering brain ac-tivity through digital EEG during the production of Vibrational States and partial projections.

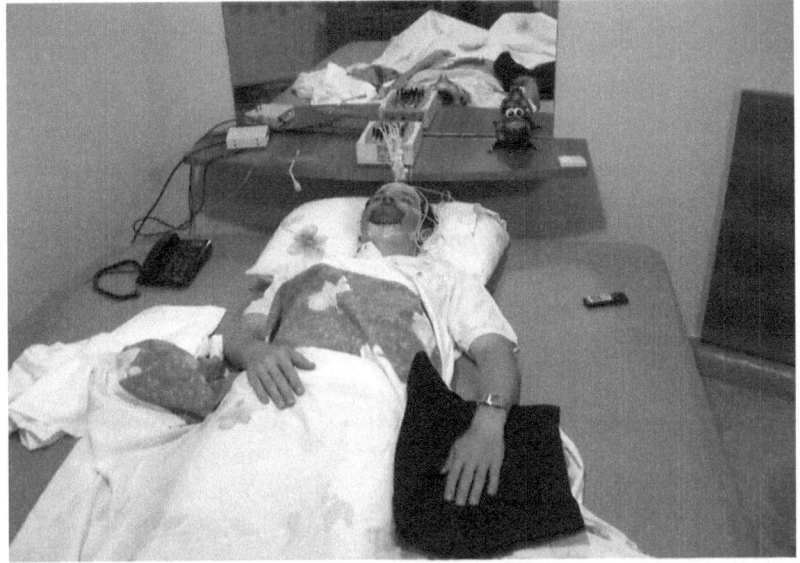

Figure 1 The author in 2007, in a typical session of digital EEG analyses, at a neuroscience laboratory in the city of Natal, Rio Grande do Norte, Brazil.

Another factor that motivated the continuation of this line of study is the research done through the bioenergetic evaluations of students in the course *Goal: Intrusionlessness*, given by this author in partnership with Nanci Trivellato since 2003, at the International Academy of Consciousness (IAC). Those evaluations brought up observations, findings and further questionings about the Vibrational State, generating new hypothesis and allowing for the perfecting of the research protocol of the research presented in this paper.

During a conference given in Belo Horizonte, Minas Gerais, Brazil, in August 2008, during the IV International Congress of Conscientiology and Projectiology, this author had the opportunity to make original comments about the investigative possibilities of such protocol, as well as about some informal results of this research available up to that moment.

Objectives of the present paper

This work aims to propose an executable and replicable protocol for the research of the VS, according to the technology and knowledge available nowadays. Furthermore, it aims at establish-

21

ing hypotheses, theories and possible future applications for this line of study.

The possible success of this approach may open doors for future similar studies, by showing that the research of certain consciential phenomena, until now considered subjective or beyond the scope of objective analysis is doable and viable through conventional physical methods and techniques. Besides proofing this methodological line, such demonstration of executability and viability would certainly stimulate studies of other consciential and parapsychic phenomena (or not purely physical) which are even more complex.

The data and preliminary informal qualitative results of the 2 experimental sessions already conducted by the author and above mentioned, which point towards the validity of the hypothesis and viability of this research and of the protocol already presented will also be discussed here.

The last section of this article presents an essay about the possible useful applications originating from the broader knowledge of the vibrational state and from perfecting the research techniques presented here, including those for the application towards other consciential phenomena, being those more closely related to bio-energies or not.

2 Basis of the research

Hypothesis

Based on a multidimensional consciousness paradigm, this research is based on three basic hypotheses:

1. Bioenergy is real and objective;
2. The VS is an objective occurrence, not being just subjective, imagination, illusion or sensory hallucination by the practitioner;
3. The VS is accompanied by detectable changes in the human brain or can cause alterations in such (some temporary and, perhaps, others more permanent).

22

Based on the specific knowledge existent today about the Vibrational State, which is still relatively limited due to the lack of systematic research about this phenomena conducted until this moment, it is not yet known whether there are true VSs, of high intensity, that do not produce any level of repercussion on the soma, or even if all the intense VSs will cause a repercussion on the soma.

Within the uncountable types of repercussions of the VS experimented, it is supposed that at least some VSs will produce a bigger effect on the soma, while others will concentrate its effects more directly on the energosoma, or maybe, on the vehicles more subtle than this one. This way, it is anticipated that possibly there are VSs that do not produce any somatic effect (or, more probably, produce a very low somatic effect) that can be registered by technical apparatus of physiological or neurological monitoring existent nowadays.

So, this research focuses on studying the VSs which effects reach the physical body, which, as previous research indicates, could be the majority, considering that the VS happens primarily in the energosoma, and that this energetic body acts as an interface between the consciousness and the physical body.

Another aspect that suggests that the occurrences of somatic repercussion of the VS are a common condition is the observation (through personal experience of this author and other peoples publications) that the great majority of the VSs experienced by the consciousness when in coincidence with the physical body are felt also on the body, or at least as physical sensations (probably, for the less sensitive individuals towards energies, those would be felt mainly *on* and *by* the physical body). This fact points to the logical assumption that, since the sensations and – or at least some – bioenergy effects manifest frequently (and at times intensively) on the soma, the VS probably also produces changes on the soma that can be detected.

Possible benefits

The experimental development of studies in this area has sufficient merit per se, but also because it would allow for promoting the following possible relevant results and discoveries, among others:

o Identification, categorization and cataloging of the neurological effects or correlates provoked by the VS or concomitant to it (relative to the hypothesis 3 above);

o Characterization of the VS as a distinct state of other neurological or consciential states;

o Collection of data and findings for a better comprehension of the VS itself;

o Demonstration of the VS as a real and objective energetic phenomenon (relative to hypothesis 2);

o Better understanding of the processes and factors involved in the development and effective willful production of the VS, allowing for the generation of more effective pedagogical methods and more exact descriptions, capable of promoting better energetic self-control for the population of practitioners of the VS technique;

o Better comprehension of some of the mechanisms of the interface energosoma-soma and parabrain-brain *(hard problem of consciousness)*;

o Gathering of a larger number of evidences that support the theory of objectivity of the bioenergy, stimulating new areas of research (including several interdisciplinary ones) and the deepening of the theoretical and practical study of consciousness (relative to hypothesis 1);

o Development of new practical applications for the VS, including therapeutic ones.

This research will also help in the tabulation (through comparisons and multiple analyses) of the classification of the VS according to the level of effect on the soma, and as a consequence, according to the types and intensity of the repercussion on the energosoma and other vehicles of manifestation, among other criteria not yet anticipated.

Once there is advancement in the comprehension of the characteristics of the VS and its taxonomy, we will be able to foresee other forms of study and verification, more detailed and directed, according to each specific type of VS.

Reasons for choosing the Vibration State

Consciousness study encompasses a wide spectrum of multidimensional consciential parapsychic phenomena (known in some areas as paranormal, psychic or spiritual phenomena). So, the choice of the VS as the main subject of study in this investigation is due to the following factors:

1. **Universality.** Even though still relatively uncommon, the VS is a human phenomenon that is fairly universal, having been experimented and described by many people, independent of sex, age, nationality, level of education, religion or culture (Vieira, 1999). Furthermore, the existence of accounts of VS registered for centuries reduces significantly the validity of any hypothesis of refutation that may claim that this may be experienced out of mere 'suggestion', originated from the influence of media or reading of books on the subject. This way, the VS seems to be a real phenomenon, possessing casuistic and evidence that seal its merit to be studied seriously.

2. **Physicality.** The personal experience of the VS, not rarely, brings sensations (and probably repercussions) so evident and real that lead the inexperienced experimenters to deduce that their VS (or the effects of such) is visible to an external spectator, therefore, manifested on the body and observable through common physical vision. The universality of the fact that the experimenter having, as first reaction, the conclusion that the soma is clearly involved in the phenomena experimented, gives support to the hypothesis that the soma manifests repercussions when under the VS. Within the many physiological systems possible of being investigated, logically, the most sophisticate and complex part, the brain, seems to be the best candidate.

According to the empiric observations of this author through his personal experiences, also in consonance with the accounts of thousands of people collected during courses and conferences on these subjects given by him since 1982, and supported by the preliminary results of the international survey about the out-of-body experience (OBE) in progress since 1998 (which accumulates to this day, answers of more than 9,000 participants of several countries), it can be deduced that the VS is, possibly, one of the consciential phenomena that has more repercussion on the soma (Alegretti & Trivellato, 1999). Such finding is further supported by analyses of specialized literature. Considering those points, the detection of the somatic components of the VS would be made easier in virtue of its characteristics and types of repercussions generated.

3. **Accessibility.** Distinct from many other consciential phenomena, the VS can be learned, practiced, provoked and repeated voluntarily at any moment with relative ease, requiring nothing but practice, decision of the practitioner and application of a developed will. Such condition does not occur in the same manner with other more complex consciential phenomena, such as, for instance the cosmoconsciousness (samadhi, satori, nirvana) or precognition. This fact shows that, even though the VS is a bioenergetic phenomenon, its practical viability and replicability make it possible to establish a universe of research participants broad enough to allow for unexpected results and findings, and the obtaining of more universal patterns, as well as contributing confidence and significance to the findings.

4. **Paraphysiology.** The deeper and more detailed knowledge about the effects of the VS can demonstrate and clarify the mechanisms of the 'first link of the chain physical-extraphysical': the soma-energosoma interface (the following interfaces or links being: energosoma-psychosoma and psychosoma-mentalsoma). Such knowledge will contribute to reach a deeper understanding of the relationship between the vehicles of manifestation of the consciousness, its mechanisms of coexistence and its respective physiologies. This information can, therefore, provide valuable insights about how the consciousness interacts with the soma and controls it, providing self-evolutionary subsidies, since the conscious control of the bioenergy-soma interaction is central in the mechanism of lucid manifestation of the consciousness.

3 The vibrational state (VS) of the research

For a better comprehension of the points discussed in this work, especially for those without further knowledge about the subject or without practical experience with the VS, it is worth presenting a brief explanation about this phenomenon. This will also contribute to the better understanding of the basic criteria of the research protocol presented ahead.

Characterization

The VS is usually described by those that experience it as a strong and exotic vibration (non-mechanic), throughout the whole body, which is more frequently associated with sleep, namely the hypnagogic or hypnopompic phases and the occurrences of out-of-body experiences.

One characteristic element of the accounts of the VS is the description of a sensation as if the cells, molecules and atoms of the body were at a high level of frenetic oscillation, but cohesive as a whole or in unison. Included in the layman expressions more frequently used to describe the VS are: pleasant vibration; generalized tingling; painless electrical shocking, internal strength or power; roaring waves; internal static electricity; internal chills, energetic incandescence; a dazzling aura, effervescence; super-vitality; etc.

In some cases, the VS happens associated to phosphenes, *acouphenes* and even the projective catalepsy. Keeping in mind that the greatest majority of VSs experienced by people that did not know of it occurs in a spontaneous fashion, many get surprised by it, and some report, in some cases, feeling fear in relation to the occurrence.

Types

Besides its spontaneous type, the VS can be provoked by several factors (conscious projection, proximity of an extraphysical consciousness, action of intense immanent energies over subjects predisposed to the phenomena, deep relaxation, certain forms of meditation, hyperventilation, among others).

The VS can also occur in a voluntary fashion, generated through various techniques, such as certain forms of breathing exercise (among which is the *bhastrika pranayama* of yoga) and specially through the energy technique, classically called closed circulation of energies (or closed mobilization of energies) (Vieira, 1999), named by Trivellato, in a more objective and unequivocal way as voluntary energetic longitudinal oscillation (VELO[1]).

VSs can vary with regards to their intensity, broadness over the body, degree of subtleness or crudity, vehicles of manifestation of the consciousness involved, among other criteria.

Pseudo-VS

Often, due to the unsatisfactory knowledge of the technique and its mechanisms, as well as anxiety, laziness and self-deception, many practitioners develop approaches that lead to states that do not align with the characteristics of the real VS.

Through personal experience of this author and his lectures in this field since 1987, it has been possible to catalogue different forms of pseudo-VSs. However, the *tensional state* can be highlighted: when the person basically contracts their musculature, to a greater or lesser degree, until they feel a form of vibration on the body, an internal agitation or subtle heat. It is very common, in some cases, for the occurrence of an almost instinctive contraction of the perineum. There are occurrences when the person reaches the 'trepidational' state, when a witness can observe the occurrence of tremors, muscle contractions and spasms (*myoclonus*).

In other situations (unfortunately not rare), the person looks for shortcuts, or ways to 'speed up' the production of the VS, allowing for an unacceptable level of reduction of the quality and effects of the technique. For instance, there are cases in which with only a few seconds of certain concentration and a few deep breaths the practitioner announces having reached the VS, when, all he did was to focus on himself and become more aware of his common physiological and, perhaps, low level bioenergetic sensations.

Technique

The basic VELO can be described in a direct and didactic manner, as following:

o Remaining preferably with the body straight (standing or lying), create and move slowly a wave of energy, using a strong will, from head to toes (limiting poles). Do not use visualiza-

---

[1] Voluntary Energetic Longitudinal Oscillation – introduced by Trivellato (2008) during the *2nd International Symposium on Conscientiological Research* that took place in the IAC Campus in Portugal, in October 2008.

28

tion or imagination. Note: in case the body is in the seating position, or in other non-straight ones, the energy should accompany the position or shape of the soma, assuring its passage through the 'interior of the body'.

o   As soon as the wave of energy, permanently under the resolute command of your will, reaches your feet, invert the direction immediately, moving it upwards until it reaches the top of your head, always tracking it with attention and acuity of perception throughout the whole path.

o   When the wave reaches your head, invert the direction once again, now moving it downwards.

o   Maintain this simple, harmonic and repetitive movement, with the energy wave permanently going up and down, always under the command of your focused and determined will. The movement should be slow, continuous and smooth, meaning, no leaps or skips, or abrupt changes.

o   During the whole time, try to perceive the complete path, detecting energetic blockages, which are more commonly perceived as energetically 'dead' areas (with no sensations) or resistant to the flow of energy. Try always to overcome those blockages by forcing the energy to go through those areas.

o   Treat the whole procedure as a *true movement of real energy*, even if, in the beginning, you are not able to perceive it. The ability to feel it will come with time.

o   Always avoiding any use of imagination or visualization, try to reach a stable and clear regimen of mobilization of energies. Maintain it for some time. In the meanwhile, try to intensify, gradually, the energetic wave (in other words, to make it more vigorous). Avoid at all costs, unnecessary physical movements, tensions or muscular contractions.

o   Then, aim to accelerate the wave gradually. It is common, as you speed up the wave, to feel a reduction of the intensity of the moving energy, or the loss of rhythm or cadency. If this happens, reinitiate the whole process, from the start. Repeat as many times as necessary or for as long as your available time allows. With time you will be able to reach greater speed and intensity, without losing the rhythm, coordination or synchronism of the movement.

o   When you are able to accelerate and intensify sufficiently, you will begin to feel bursts of vibrations throughout the body, very brief and disperse at first, which gradually will become more cohesive, broad and intense. Reaching this point might take weeks or a few months for most of the people.

o You will then, one day, reach the point where you feel a sudden chain reaction of intense vibration, many times self-sustaining, that takes over the whole body. At this point, result of an exponential intensification or explosion of energy, you will perceive your body like an engine, or generator, vibrating and buzzing as with an 'electrical current of millions of volts'.

Note: The VS, when adequately produced, does not increase the heart rate, does not increase the blood pressure, does not cause spasmodic muscle contractions, does not raise the body temperature nor makes the practitioner blush.

It is necessary to highlight that it usually takes a few years of daily practice until you are able to reach complete, intense vibrational states, at any time, and any place, independent of any factor (internal or external) other than your determined will.

Conceptualization

Even though we still know very little about the physiological and paraphysiological mechanisms of the vibration state (a gap that we intend to fill at least partially with this study), the VS is understood as a resonance, expansion or profound and intense activation of all the structure of the holosoma (set of the bodies used by the consciousness, including the physical one) and its energies. In the case of the intraphysical consciousness, the activation of the energosoma predominates, which includes meridian (*nadis*), energy points (acupuncture points) and chakras. In the case of the extraphysical consciousness, the activation of the psychosoma prevails.

Such resonance seems to cause or to come from an alignment or coherence of several waves, systems and natural regimens of oscillation of bioenergies. Many times, the resonance promotes, or at least facilitates, the looseness of the psychosoma, being able to provoke conscious projections. At other times the VS seems to be the result of a certain level of disconnection of the vehicles of manifestation of the consciousness.

The experience and cases studied show that some people have a greater predisposition than most to have spontaneous vibrational states. In the case of the provoked VSs, some people require relatively little time to master the dynamics of its generation, while the majority has difficulty provoking it or experiencing it.

Under the conscientiality's point of view (Vieira, 1999), it seems that this variation on the degree of susceptibility or natural predisposition to the occurrence of the VS is due to the consciousness having or not passed through the second desoma (the deactivation of the energosoma) during his/her last few intermissive periods (the periods between lives), as well as the quality of his/her so-called thosenes (the practical unit of manifestation of the consciousness: *tho*ughts + *sen*timents + *en*ergy), relative degree of intrusionlessness among other factors. The research proposed in this work will maybe allow for a better understanding of the physiological factors involved in this degree of energetic susceptibility.

Analogy

The regimen of working, the mechanism and genesis of the VS, greatly resemble the concepts and functioning of a resonating cavity, and more specifically, of the laser. Especially when we take into consideration the VELO technique described above, the VS resembles the classic ruby cylindrical crystalline laser, with two mirrored extremities (one totally and the other one partially reflective), in which light reflects and propagates uncountable times within the crystal, stimulating the production of more light in the same frequency, phase and plane of polarization, creating what is called coherent light.

Analogically, in the case of the execution of the VELO, the cyclic voluntary movement of the energy up and down within the body seems to stimulate even more the 'liberation' (release) of the bioenergy. With the frequency increase (increasing the up and down speed) combined with the increment of the amount of free energy in movement, a sort of chain reaction is triggered, many times self-sustaining, which is felt by the experimenter as the VS.

Following this same analogy, it is unknown in the case of the VS, about the mechanism equivalent to the *pumping* of the laser. Probably it would be some form of action of the energies of the mentalsoma.

Parabrain

When assuming that the VS is an objective state or energetic regimen that happens, mostly, in the energosoma, we get to the logical assumption that it cannot be provoked or controlled only by the physical brain (which has its action restricted to the physical body).

It's supposed then that the parabrain is the effective center of the initial command of such consciential action. This supposition is supported by the fact that it's possible to install the VS even during a conscious projection of the psychosoma, meaning, without the presence of the body or physical brain.

In the model of a multidimensional consciousness it's assumed that every voluntary action originates from the consciousness, passing through the mentalsoma and through the parabrain. This way, in the case of the physical actions, the command coming from the parabrain is received, translated and adapted to the animal physiology, mediated by the energosoma, resulting in the somatic action.

In the case of the VS, the most part of the command originating in the parabrain reaches (or should reach) only to the energosoma, producing then energosomatic actions or bioenergetic effects.

The central hypothesis of this work is that a certain percentage of this command of the parabrain or the energies that are activated through the vibrational state echo in the physical brain. As previously discussed, given the fact that we can feel the VS physically and that we create it while within the physical body, there should be a 'neurological echo' or correlate associated to it.

Parapsychomotricity

When analyzing the VELO technique attentively we see that, even though it is simple in its procedures, it requires the development of a specific form of motor coordination, or, more appropriately, a para-motor coordination or , better saying, parapsychomotricity, as it involves the parabrain and voluntary actions of the consciousness over the subtle vehicles (beyond the soma).

The complexity comes from the need for concomitant and synergistic application of its spatial directives (direction, sense and completeness), temporal (frequency, rhythm, synchronization and acceleration) and energetic (perception of the energy for the necessary feedback of control, increase of the amount of energy, overcoming of energetic blockages, incrementing the fluidity of the energy, and increasing the multidimensional depth).

As in any type of psychomotricity, the development of such capacity requires time, repetition and practice.

Benefits

Among the benefits observed with the practice of the VS, be it immediate or over time, we may highlight: energetic unblocking in general; intensification of the energies of the energosoma; relief or cure of diseases or health problems in a greater or lesser degree; activation and development of chakras; increase of the degree of parapsychism; stimulus and facilitation of the projectability, including the lucid take-off; expansion of the level of energosomatic vigor; auric decoupling; sympathetic de-assimilation; energetic self-hygiene; improvement of thosenic immunity and prophylaxis; a more stable and permanent energetic self-defense; and an increase in the level of intrusionlessness.

As far as the uses of the VS beyond the manifestation in the physical dimension, it includes the possibility of the consciousness, when projected with the psychosoma or when extraphysical, utilizing the extraphysical VS as a resource for changing dimensions.

4 Methodology

The somatic and inter-vehicular repercussions of the VS could be studied according to different criteria. The ideal would be through a bioenergetic technology that would allow for the detection and direct measurement of the energies of the VS. However, obviously, this technology does not yet exist.

So, to allow for the realization of a more objective exam of the effects of the VS and the systematic comparison of the results, the most effective and consistent method seems to be the detection of the neurophysiological alterations of the practitioner through the technology for analysis of neurological functions available today. This method also allows for the replicability of the research by an independent researcher (including those who have never felt or produced a vibrational state).

This way, this study protocol will use investigative resources common to laboratories of neurological analyses, with special emphasis on the EEG and fMRI.

It is worth mentioning that there are other forms of study of the VS and of the vehicular interface of the consciousness, but those are less objective in nature.

The possibility of making a comparative analyses (*cross-analysis*) of the final results of this present study with the results of other similar researches (with the same objective, however, with different methodology, such as for instance, the research conducted with the direct sensing of someone's VELO and VS) might be interesting. Such comparative analyses will allow for the validation of the methods applied and the results of the research, creating, as a consequence, the generation of new research hypothesis and planning of different lines of investigation.

Experimental Techniques for Data Collection.

Among the several resources available today for the real time analyses of brain functioning, two techniques were selected due to their characteristics. Keeping in mind the pros and cons of such techniques, the ideal is that both be used in a complementary manner.

1. EEG (Electroencephalography or Electroencephalogram)

The most simple, accessible and affordable method that provides better temporal resolution of the brain's dynamic. It consists of the recording, almost always at the level of the scalp, of the electrical activity of the brain, or more precisely, of the summation of the electrical potential of large groups of neurons.

As an older technique, it allows for the comparison of new findings with several other studies already exhaustively accumulated, analyzed, catalogued, studied and characterized for several decades (e.g., patterns of brain waves, cerebral arrhythmias, epilepsy, sleep studies, etc.).

When compared to the fMRI, and under the point of view of the subject being studied, it counts with the following advantages: less sensibility to physical moments; less noise; and lesser incidence of claustrophobic events – facts that could introduce several non-controllable variables to the experiment.

With the current availability of the digital version of the EEG, the recording, storage and later treatment and data analyses, through specialized software, became much easier and deeper.

2. fMRI (Functional Magnetic Resonance Imaging)

A more modern, expensive and less accessible method, that presents however, better spatial resolution. Allows for the more direct, precise and immediate localization of cerebral areas involved in processes and phenomena under study. It allows, therefore, for the visual observation, in a dynamic manner, of the neurologic occurrences of several functions and cephalic areas.

As an additional advantage when compared to the EEG, it eliminates the discomfort of having several electrodes glued to the skin and scalp.

It presents the technical disadvantages of being more prone to artifacts, very low signal-to-noise ratio and still subject to problems with the statistical approaches and techniques for the data analyses.

Experiment Protocol

In accordance with the above discussion, the protocol here presented applies both to the EEG and fMRI, since they both can be used as data collection techniques.

The technical procedure for the attainment and installation of the VS (through the VELO technique) presents a certain complexity in terms of the several mental commands applied and its synchrony, when compared to other simpler actions such as moving a finger or seeing a colored light. For this reason, when we take into consideration also all the procedural vices and bad habits manifested by many practitioners of the VS, we saw the necessity of establishing a protocol with strict criteria that allows for the isolation, and later removal, of those interferences of the 'pure' VS. Otherwise, such factors could generate spurious effects in the results of the research.

The protocol presented hereinafter, is organized in phases and stages that aim for the isolation of the execution of the technique to obtain the VS, of its practical detours and also the VS as a phenomenon in itself. This phases and stages should be applied regardless of the study technique adopted (EEG or fMRI).

What is proposed therefore, is the recording of the cerebral dynamics of the participant-subject for the several 'modes' of operation listed on the phases and stages bellow, which should be done voluntarily by the practitioner, aiming to allow its measurement for future studies. Each of those stages or modes of operation should be analyzed with an equal degree of detailing and acuity.

The recording and analysis of the brain activity in certain stages described below (apparently disconnected of the objective of the experiment) have the purpose of working as a 'control' reference for the experiment, since they will allow the comparison of the results with the records obtained from the voluntary and correct application of the VELO and possible installation of the VS with great intensity.

The practical procedures, or 'operation modes', to be executed by the participant-subject are divided in the following phases, with its respective stages.

Phase 1

General Objectives: To obtain first findings; to characterize neurologically the VS to direct the next phases; to perfect procedure and techniques of data collection; to establish an universe of possibilities; to test hypotheses.

Subjects: Individuals possessing a high degree of control of the VS, carefully chosen.

Universe: 5 subjects.

Strategy: 5 initial sessions of data collection (on different days, and if possible, far apart) and exhaustive analyses of the results for each subject. Repetition of the experiment according to the results, needs or deviations encountered.

Stages:

1. *Only* production of the state of mental and somatic relaxation;
2. *Only* execution of rhythmic inhaling and exhaling, conscious and voluntary (slow in the beginning and then with its gradual acceleration);
3. *Only* execution of eye movements up and down (slow in the beginning and then with its gradual acceleration);

4. *Only* visualization of the movement of energy, meaning, without actually moving any energies voluntarily (slow in the beginning and then with its gradual acceleration);

5. *Only* sweeping of attention and perceptive focus through the soma (slow in the beginning and then with its gradual acceleration). In this case, the practitioner will try to concentrate exclusively on the existence of the part of the body that is being focused on. Such focus will move continually and cyclically up and down along the body (from feet to head and vice-versa);

6. *Only* mobilization of energies, meaning, exert smooth action towards the mobilization of energies without reaching the level of energosomatic activation that generates the VS (the energies should be mobilized slowly in the beginning and then with gradual acceleration);

7. Effective installation of the VS through the correct and vigorous application of the VELO.

The stages 1 through 6 of the experiments and analyses above described make possible for the accomplishment of the basic strategy, which aims to subtract from the data set relative to *Stage 7* the signals obtained in the previous stages, determining in this way the profile of the VS itself, separating its neurologic 'signature' of the other associated neurological processes, be them natural or derived from the technique application.

Even though the procedure described on *Stage 1* is unnecessary for the production of the VS (though not counterproductive), its recording and study are essential to establish the *baseline*, meaning, the specific basic resting neurological condition specific and particular to each participant. This resting condition will be an important reference for posterior comparisons and analyses.

The steps described in the *Stages 2 to 6* aim at simulating a pseudo-execution or partial execution of the technique of installation of the VS, having been inserted to take into consideration also the habits (some of them inappropriate) common in the application of the technique. The arguments for this insertion are better detailed below. Evidently, the data obtained during those stages have the objective of being more than 'removable noise', since the careful analyses of those can lead us to a better understanding of the mechanisms of the technique of the VELO and of the *pros* and *cons* factors to the attainment of the VS. Furthermore, they will allow for a clearer verification of the influence or not of those somatic or mental procedures over the VELO and the VS.

More specifically, the actions on the *Stages 2* and *3* (more mechanical) are undesirable for the correct and efficient installation of the VS. However, since they are extremely common among the new and several veteran VS practitioners, they should be recorded, for its subtraction from the data set obtained during *Stage 6*, as well as, when applicable, for the analytical comparison with the procedure considered correct.

The procedures designated to *Stages 4* and *5* (more mental) are possibly inseparable from the correct procedure, however more complex, performed during *Stage 6*. The knowledge of the brain's behavior during the execution of those stages will allow for the identification of what really happens, in bioenergetic terms, during *Stages 6* and *7*. This will be done through the comparison of those signals and also of their withdrawal from the data set in its last stages *(6 and 7)*. Another aspect is the study of how much *Stages 4* and *5* participate in the correct procedure for different subjects.

As in many experiments of this type, the use of an intercom between the experiment room and the room where the researchers and the equipment are would facilitate and enrich many of the results, for allowing, among other possibilities, that the subject announces to the researchers beforehand of what he wants to do, or what according to him just happened or, yet, what is taking place at that exact moment.

The use of an audio signal that can be followed by the practitioner when executing *Stages 2* to *6*, that can also be introduced in one of the channels of data acquisition (specially, with the EEG, due to its better temporal resolution), will allow for better precision in the temporal analysis and better synchronization in data comparison. This audio signal would be used as a kind of metronome, with the purpose of synchronizing the movement of energies. For instance, a pure tone would have its frequency increased to a higher pitch (that would be associated by the practitioner with a movement of energies upwards) and then the frequency would decrease to its original pitch (associated with the movement of energies downwards). In little time the practitioner would adjust its

VELO to have the energy at the head when reaching the higher tone, and to have the energy at the feet when reaching the lower tone. The time of sweeping would be programmed to be reduced (increasing the speed of the VELO) according to an acceleration that is compatible with the parapsychomotricity of the subject.

This resource will allow, for example, among other things, for the following possibilities:

o To verify what happens (neurologically) at the exact moment of the inversion of the energetic movement (i.e. at the head/coronochakra or feet/solechakra).

o To compare the records of the bioenergetic movement 'upwards' with the 'downwards' movement. Note: some practitioners perceive easier or more clearly the energies when those are moved in a certain direction.

o To observe the change in intensity of certain signals in accordance with the speed of the bioenergetic movement. Note: it's common for practitioners to describe a decrease in energetic sensation (energy intensity) as the speed increases.

o To determine the speed in which the practitioner 'loses' the synchrony of the coordination of the rhythm along the acceleration of the VELO and observe the neurologic effects of such asynchrony.

o To discover the approximate localization of probable energetic blockages, through the observation of when the signal (especially of the EEG, due to its better temporal resolution) presents abrupt changes (compared to the more stable regimen until that moment).

For the initial characterization of the true VS and the correct application of the VELO technique, expected during *Phase 1*, participants will be selected, according to interviews and measurements done by capacitated bioenergetic evaluators, who can effectively generate a VS through their bioenergetic self-control and developed will. This will save time during the experiments and will avoid the accumulation of spurious data that may come to confuse posterior analyses.

After the characterization of the profile of the real VS and VELO, analyses should be done with other people who simply know the VELO technique and consider themselves able to produce a

VS, meaning, not necessarily individuals that have a greater ability or control over the VELO and VS. Those data will also serve as comparison and control for the research.

Phase 2

General objective: to test and expand the *Phase 1* findings; to find and establish neurological and somatic patterns characteristic of the VS.

Subjects: individuals with a good level of control of the VS. Include subjects predisposed to the lucid projection.

Universe: 10 subjects.

Strategy: recording and study of self-induced VSs in 2 sessions (during distanced, different days) for each subject. This study should be done through the tabulation of the results of the experimental sessions with each person, aiming also, for the comparison among results of different people and also with the results obtained during *Phase 1*. For the study of the spontaneous VSs detailed bellow, there will be required as many sessions as found necessary.

Stages:

1 to 7. Repeat the *Stages 1* to *7* as described above;

8. Analysis of the spontaneous VSs associated with the lucid projection. In this case, the subject should relax deeply and try to predispose himself to the lucid projection. When and if it happens, the subject should notify the researchers through the intercom or press a button that will allow for the registration of that moment in the collected data. Several sessions will probably be necessary for each participant, until the spontaneous VS associated with the lucid projection can occur. Note: 1. Only in this case this specific analysis can be performed, being that the spontaneous VS as well as the self-provoked, previously registered VSs, should be compared; 2. The projective technique should be carefully chosen as not to cause any interference or contamination of the data (example: not using the rhythmic breathing technique).

Phase 3

General Objective: to generalize patterns for the VS to the point of recognizing more universal and representative patterns for human beings. Make better evaluation of alternative techniques to reach the VS.

Subjects: familiar with the VELO that consider themselves capable of producing a VS.

Universe: 35 subjects.

Strategy: one session of data collection for each of the subjects. Make comparative analysis with the findings of *Stages 1* and *2*.

Stages:

1 to 7. Repeat *Stages 1* to *7* described above;

9. Study of the installation of the VS through alternative techniques (other than the VELO), with posterior comparison of the results with those of the VS obtained through the VELO, those of this group as well as the previous ones.

Phase 4

General Objective: detection and study of the more permanent effects and neurological changes caused by the VS.

Subjects: previously selected (random sample); plus people without any experience with the VS, being it spontaneous or provoked.

Universe: 10 subjects.

Strategy: Long term follow up (longitudinal study), through some sessions of data collection with each one of the subjects. The subjects should maintain, ideally, the daily training of the VS during the whole duration of this period of research. Support the experimental research with interviews and questionnaires for data collection and search for variables that may interfere with the VS (health, work, affective life, practice of sports, medicine use, etc.).

Periodicity: annual.

Duration: 5 years.

Stages:

1 to 7. Execute the *Stages 1* to *7* described above.

Questions and phenomenological aspects to be investigated

Even though associations, theories, projects of future experimentation, methodology adjustments, not foreseen here, should appear after the final conclusions (or even after each of the described phases) of this research, it is worth remembering that the scope of study presented in this article is very broad, including several topics and research questions, some of them exemplified below:

1. Which types of brain waves or specific electrical correlates are associated to the VS? Will the association with the gamma waves be confirmed? Will the synchronizations between different areas of the brain be confirmed?

2. Would the electrical correlates associated with the VS be from the brain as a whole or from specific regions?

3. Which are the most active brain areas during a VS? What is the degree of specificity in this area?

4. What are the other functions of this area? Among them, would there be any area considered to have no specific function until then? Would the remaining functions of such areas have any direct or indirect relation with the VS?

5. Would the VS be associated with a specific cerebral correlate or state, meaning, an unknown state or not yet associated with previously catalogued condition?

6. Would there be a participation of the anterior cingulate cortex (anterior cingulate gyrus)?
   Dr. Olaf Blanke observed, during an epilepsy corrective surgery, that the electrical stimulation of this area provoked sensations normally related to the OBE in his patient (Blanke, 2004). The anterior cingulated cortex (ACC) is a complex part of the brain, rich in spindle neurons, possibly responsible for processes such as self-consciousness in regard to body and space, as well as logical sequencing. Physiological and anatomical abnormalities of the ACC are connected to the most important psychiatric disturbances such as autism and ADD.

7. What is the importance of the fronto-insular cortex in the VS? This area, also rich in spindle neurons, seems to be associated to functions such as: consciousness of emotions and sentiments; translation of corporal sensations as emotions; integration of internal sensations to superior cognitive functions; and empathy.

8. What is the role of the spindle neurons (*von Economo* neurons) in the VS?
   The spindle neurons are more abundant in the anterior cingulated cortex and also in the fronto-insular cortex. They seem to

have the function of integrating and connecting relatively distant areas of the brain. It was once thought that those neurons existed only in human beings, bonobos, chimpanzees, gorillas and orangutans (listed here in decreased order of abundance of those neurons). However, recently they were also discovered in elephants and certain cetacean, precipitating the hypothesis of them being related with the manifestation of superior expressions of intelligence and self-consciousness. According to Dr. John Allman (2001, 2002, 2005), they seem to be involved with the expression of attributes such as will, self-control, decision making, and discernment. Anomalies in the development or degeneration of the spindle neurons seem to be associated with Alzheimer's disease, fronto-temporal dementia and also many of the psychosis, including schizophrenia.

9. What is the importance of the participation of proprioception?
10. What is the importance of the interoceptive senses?
11. What is the importance of the right temporal lobe?

    Dr. Melvin Morse speculates that the right temporal lobe would have a special connection with near-death experiences and other transcendent, mystical and religious experiences (Morse, 2008).

12. Would there be any specific relation with the glial cells?

    Seen in the past as cells with simple structural function, immunological and for chemical homeostasis and nutrition of the neurons, nowadays they are known to have more complex functions such as in the participation of synaptic plasticity. Some forms of glial cells have synapses and produce neurotransmitters, while others produce calcium waves, which could have some function in the intra-cerebral communication. Note: in reality, it is change in blood flow (hemodynamics) in different areas of the brain, mediated by a specific type of glia (the astrocyte), that is measured by the fMRI.

13. How can we know, physiologically, if the VS happens only in the brain or in the whole body?
14. Is there the mediation through a specific neurotransmitter?

It is probable that new lines of research, hypothesis and investigative processes will arise as a result of the discoveries above mentioned, and others, not foreseen at this moment.

5 Preliminary experimental results

To illustrate and also to reinforce the viability of the hypotheses adopted and the methodologies proposed for this study, it is worth to briefly present here some data about the preliminary experiments already conducted by this author in this field of research.

Santa Casa Hospital, Sleep Study Laboratory, Porto Alegre

In 1991, this author was invited by one of the doctors of the Sleep Study Laboratory of the Santa Casa Hospital, in the city of Porto Alegre, RS, Brazil, to participate in experiments aiming to study the physiology of the lucid projection.

In a session of approximately 2 hours, the author was laying in the dorsal decubitus position on a bed in an isolated room, connected to EEG, ECG equipment, breathing rate monitor, electromyograph to register MRO activity and oxygen saturation monitor. The system was analog, with a mechanical polygraph, so that it was registering in paper the signals received from each of the respective circuits described above.

The author executed several experiments, such as: deep relaxation; to enter, stay and leave, at will, the hypnagogic state (alpha); and attempts at the lucid projection and maneuvers of basic mobilization of energies (absorption, exteriorization and VELO aiming to reach the VS). At the end of the experiment, in discussions and joint analyses with the physician responsible, we observed the occurrence of the facts listed below, relative to the closed mobilization of energies and attempt of lucid projection:

o   The synchronization of several brain circuits, considered atypical by the doctor, occurred during the installation of the vibrational state.

o   Waves of greater frequency superimposed over others (alpha and theta).

o   Incomplete cycles of certain waves, out of the median line, presented for instance, only the positive semi cycle.

o Similarity with regimen of typical waves of cerebral arrhythmia. Such record motivated the doctor to question if this author had ever had epilepsy or similar conditions (fact that never occurred).

According to previous agreements with such doctor, the studies and more in-depth analyses, such as copies of the records of the polygraph, would be shared with this author. However, unfortunately, this author was never able to obtain it. Such fact impeded a more in-depth analyses of the occurrences and also prevents the inclusion here of greater technical and precise details about such experiment, affecting the possibility of a more detailed comparison of that experiment with others.

International Institute of Neuroscience of Natal

In December 2007, the author and Nanci Trivellato were invited by the IINN – International Institute of Neuroscience of Natal (city of Natal, RN, Brazil) to participate as subjects of research in experiments about lucid dreams.

During previous conversations about the experiments, this author proposed to the chief researcher, Dr. Sidarta Ribeiro, that we took advantage of the opportunity to also do experiments on the VS, according to the protocol similar to the one presented by the author in the *Stages 1* to *7* described above. The goal and initial proposal were to utilize the fMRI equipment for such sessions. However, due to technical problems, it was only possible to do the experiments through the computerized digital EEG, installed at the IINN.

Before monitoring the experiments related to the lucid dream, the experiments related to the VS were executed. The fact that the experimental session, of which this author was the subject, lasted the whole night, allowed also that he focused on the lucid projection, aiming to have the brain activity involved in the different stages of this phenomena also registered.

Even though this author was not successful, in that occasion, in the production of lucid dreams or a complete out of body experience (only semi-projections occurred), he was able to reach a few voluntary VSs of good intensity and one spontaneous, pre-projective VS of significant magnitude.

In the following morning, the informal preliminary analyses

45

done by the several members of Dr. Sidarta team pointed to the following facts, worth noting:

o   The synchronization of several brain circuits;
o   The appearance of atypical waves of high frequency (gamma waves);
o   Wave forms surprisingly different, since there were no known activities that could have produced such pattern of brain waves.

Despite the lack of a greater technical rigor in this previous evaluation (also because the analyses of that data have not yet been concluded), several of the researchers commented, with a certain surprise and scientific curiosity, that they "had never seen a brain functioning in that manner".

Such findings — obviously still preliminary and without the desired scientific rigor — reinforce the validity of the assumed hypothesis and stimulate the motivation to continue with this line of research. Maybe it will also be possible to confirm that a lucid projection "is produced by the increase in vibrations of the vehicles of manifestation of the consciousness, including here the human body and the mentalsoma" (Projectiology, 1999, pg. 205).

6 Possible future applications

Starting from the accumulation of data, the expansion of collection of case studies, the widening of the knowledge about the phenomena, and especially the establishment of average values and behaviors of the VS through the examination of the greatest number possible of participants, it will probably be possible to develop the following practical applications, among several not yet glimpsed:

1.   **Independent measurement:** External detection of the VS in any person, including those that are still developing their specific parapsychomotricity, and therefore are not yet confident or lucid about their experiences, diminishing the doubts about the existence or actuality of their VSs.

2. **Energometry:** Indirect estimate, through the neurophysiological measurement, of the power and broadness of the VS, allowing the practitioners an initial feedback to facilitate their development.

3. **Qualification:** Analysis of the quality of the VS, through the indirect measurement of the attributes associated with its generation, such as: quantity of energy, velocity, rhythm, broadness, cohesion, activation and others (Trivellato, 2008).

4. **Intraphysicality:** Determination of the percentage in that a specific VS is physical (or has a somatic repercussion). It is anticipated here, for example, possible cases in which the participant has a VS that operates or happens mostly in other more subtle vehicles. In such cases, the cerebral analyses could indicate weak signals in where the participant could be convinced of having experienced an intense VS, but yet, subtle. The occurrence of a true VS could be confirmed by an external agent (researcher sensitive to bioenergies, able to measure the VS and its intensity), as to confirm the existence, in this case, of a VS with less interface or action over the physical vehicle.

5. **Mechanism:** Better understanding of the mechanisms of action of the factors that intervene in the VS: positive or negative; endogenous or exogenous.

6. **Classification:** Possibility of characterization and contextualization of different types of VS.

7. **Projectability:** Possibility of detection of the imminent projection, when it is associated with the occurrence of the VS (a common condition in many out-of-body experiences). In certain cases, this detection could be used to generate the extraphysical awakening of the consciousness and to help in the obtaining of lucidity and control of the projectability (in case the VS happens during take-off), or yet to help stimulate the recollection of the projective experience (when the VS happens during the return to the soma). In other cases, it could allow for the objective and technical study of the lucid projectability by researchers of OBEs.

8. **Support:** The development of a supporting technology, in the form of biofeedback, which would facilitate the beginners to develop the ability to generate the VS.

9. **Parapedagogy:** Perfecting of the teaching method of the VS technique. All the findings obtained through this line of research would be used right from the beginning of the teaching process for the production of the VS, to improve the techniques per se and also the clarifications for their application.

10. **Therapy:** Improvement of certain therapeutic and self-therapeutic techniques based on the multidimensionality of the consciousness. Given the importance of the VS in the personal control of energies and of the parapsychism, as holosomatic homeostatic resource (physical, energetic, emotional and mental balances), and as an energetic self-defense technique, it will be evident to the application of the findings in this line of research aiming the integral health of the consciousness.

11. **Intraphysical-physical interface:** Support to the development of the *bioenergetic technology*. The detection of the VS could be one of the first steps for the detection of bioenergies (indirectly), or at least of specific regimens of those. As such, it would contribute for research and development of bioenergetic apparatus capable of interacting directly with bioenergies (detectors, transducers, meters, accumulators, transformers, *imagers*, etc.), until we reach the point where the bioenergetic technology works as an intraphysical interface with the extraphysical paratechnology.

Possible future extensions of this research

Besides the possible future application described above and even before those, the scope and methodology of this study could still be adapted and extended for:

o Neurological analyses of the VS through the use of other technologies:
  o PET scan (Positron Emission Tomography)
  o NIRS (Near Infrared Spectroscopy)
o Analyses of other effects of the VS on the soma:
  o Study of the biochemical changes: hormonal and metabolic
  o Research on the influence over the immune response
o Study of the epigenetic changes, meaning, on the pattern of

expression of certain genes (which ones would be activated; which ones would be deactivated; what would be the mechanism, etc.)
o Research of other energetic maneuvers:
    o Absorption of energy
    o Exteriorization of energy
o Research of other consciential phenomena:
    o Lucid projection (out-of-body experience)
o Extrapolation and universality:
  Study of the possible spontaneous occurrence of the VS in animals.

7 Conclusion

Considering, in 'cost x benefit' analysis, the following factors:

(1)   the accessibility and executability of the proposed methodology;

(2)   the availability of subjects;

(3)   the fact that the necessary technology is available and accessible;

(4)   the preliminary experiments already executed that support the presumed hypothesis; and

(5)   the importance and utility of the possible discoveries and expansion of the knowledge about a universal phenomenon, but not properly approached by academic science;

this author considers that not executing this line of research would mean a great loss of opportunity for the expansion of human knowledge in the direction of multidimensionality.

This research's worth is increased when taking into consideration that its relative objectivity and the acceptability by the materialistic-reductionist researchers could contribute to the arousal of the interest in those with an open mind, who, however, have more conventional perspectives about the nature of the consciousness or are more skeptical in regards to multidimensional consciousness paradigm – almost always due to the lack of personal multidimensional experiences.

Bibliography

Alegretti, Wagner (1990). Tecnologia Bioenergética (Bioenergy Technology), Proceedings of the 1st International Congress of Projectiology, IIPC, Rio de Janeiro, Brazil, p. 32.

Alegretti, W. & Trivellato, N. (1999). Pesquisa de Opinião Pública sobre a Experiência Fora do Corpo (Survey on the Out-of-Body Experience), Proceedings of the 1st International Fórum of consciousness Research and 2nd International Congress of Projectiology, Barcelona, Spain, IIPC, October 21-24, 1999.

Allman, J.M., Hakeem, A. & Watson, K. (2002). Two phylogenetic specializations in the human brain, The Neuroscientist 8(4), pp. 335-346.

Allman, J.M., Hakeem, A., Erwin, J.M., Nimchinsky, E., & Hof, P. (2001). Anterior cingulate cortex: The evolution of an interface between emotion and cognition, Annals of the New York Academy of Sciences 935, pp. 107-117.

Allman, J.M., Watson, K. K., Tetreault, N. A., & Hakeem, A. Y. (2005). Intuition and autism: a possible role for Von Economo neurons, Trends Cogn. Sci. 9(8), pp. 367-373.

Blanke, O., Landis, T., Spinelli, L., & Seek, M. (2004). Out-of-body experience and autoscopy of neurological origin, Brain magazine 127(Pt 2), pp. 243-258.

Morse, M. (2008). *Verbal communication*. Mind without brain: a scientific analysis of near-death experiences with special attention to those in children: A new scientific paradigm of consciousness, 4th International Congress of Projectiology, 15-17 August 2008, International Institute of Projectiology and Conscientiology.

Trivellato, N. (2008). *Verbal communication*. Mensurable Attributes of the Vibrational State Technique, Conference presented during the 2nd International Symposium of Conscientiological Research, October 18, 2008.

Trivellato Nanci & Alegretti, Wagner (2005). Bases para o Energograma e Despertograma, Conference presented during the: I Jornada da Despertologia; CEAEC, July 15-17, 2005, Foz do Iguaçu, Brazil.

Trivellato, N. (2017). *Vibrational State and Energy Resonance*, International Academy of Consciousness, USA.

Vieira, W. (1999). *Projeciologia: Panorama das Experiências da Consciência Fora do Corpo Humano*, Instituto Internacional de Projeciologia e Conscienciologia, Rio de Janeiro, RJ, p. 384.

Note: This article was previously published in: Journal of Consciousness 11, No. 42, 2008, pp. 217-251. An Italian translation is also available in: Studi sulla Coscienza, AutoRicerca, Numero 10, Anno 2015 (www.autoricerca.ch).

AUTORICERCA

Measurable attributes of the vibrational state technique

Nanci Trivellato

Issue 20
Year 2020
Pages 51-87

Abstract

When studying a phenomenon, establishing a measuring system that allows systemized and more objective comparisons and observations is desirable. Measuring bioenergetic and non-physical elements has proven to be a great challenge for researchers of the multidimensional reality of the consciousness and its paraphysiology. Thus, the Vibrational State (VS), one of the most fundamental resources of lucid parapsychic self-control, still awaits better studies of its modus operandi and effects. Within this line of reasoning, this paper presents a measuring system for the VS and its promoting technique, which has been used for over 5 years, referred to here as the Voluntary Energetic Longitudinal Oscillation (VELO). The discussions focus essentially on the descriptive and parametric elements of the VS (its attributes), which are identifiable also by external agents, allowing for a less subjective measure. It is anticipated that knowledge of these attributes will grant resources for a biofeedback process, which will favor the control of the willful installation of the VS.

AutoRicerca, Issue 20, Year 2020, Pages 51-87

1 Introduction

The refined control of the technique for producing the personal phenomenon known as the *vibrational state* (VS) is among the most complex aspects of basic bioenergetic procedures. Nonetheless, the effort, dedication and time invested in reaching such control are, evolutionarily speaking, highly profitable, due to its multiple positive effects.

The VS leads individuals to a level of self-knowledge of their personal energetic condition that empowers them to identify details and subtleties of their energetic body. Consequently, it allows them to discern, instantaneously and with certainty, any changes that may occur in their own energy field, whether the changes were produced by themselves or generated by another consciousness or by another form of external interference.

Having real control of the technique for producing the VS, as well as frequently experiencing it, provides the consciousness with a type of 'energetic control tower,' which leads to the development of a multiplicity of bioenergetic aptitudes; thus, bestowing on the consciousness a proficiency to understand and produce a series of other personal bioenergetic and parapsychic phenomena.

The installation of the VS, whether by the consciousness' direct control of his/her energetic interface (energy body) or originating in a spontaneous or intuitive fashion, can produce a preventative or curative bioenergetic asepsis of his/her energetic interface or aura. With time, this bioenergetic phenomenon also leads the consciousness to a more complete energetic balance, as well as to a more stable and permanent energetic self-defense and endurance.

The capacity of inducing such phenomenon willfully – in any condition, any place and at any moment, based on one's actual bioenergetic self-control and direct action over one's energetic interface – requires (1) knowledge of, (2) identification of, and (3) action upon certain key attributes involved in the application of the technique to produce the VS, which are discussed below.

The technique for inducing the VS, more commonly known as

Closed Circuit of Energies, Closed Mobilization of Energies, or *Closed Circulation of Energies* constitutes one of the most basic yet paradoxical bioenergetic procedures, since, on the one hand, it is extremely simple, but, on the other hand, it presents significant complexity to coordinate the elements involved.

Actually, this technique corresponds to a *cyclic longitudinal energetic mobilization* throughout the energetic interface or to a *longitudinal oscillation of energies,* which is performed in a willful fashion. This energetic procedure 'organizes' the spontaneous bioenergetic movements of different natures, frequencies, modes and patterns occurring in the practitioner's energetic body, transforming them into a type of coherent stationary wave that encompasses the entire energetic body.

One *session* of the technique corresponds to a continuous mobilization of an energetic pulse in complete successive longitudinal *cycles* through the length of the energetic interface. These cycles, which comprise <u>paths</u> parahead-parafeet and parafeet-parahead, occur in an uninterrupted manner. At the end of each path (i.e., in the coronochakra or solechakras) a new 'push' is applied to the energetic pulse by the consciousness by means of the application of his/her *will.* The objective of such a procedure is to reach a point in which a cohesive and stable stationary wave is created.

It is not the objective of this article to teach how to move one's energies, nor to explain what the VS is or describe its sensations or effects.[1] The purpose of this work is exclusively to state some of the main elements that need to be coordinated in order to reach an effective control of the Energetic Longitudinal Oscillation, as well as to propose a methodology for the study and measurement of the elements cited.

NOTE: To favor precision, didactics and the avoidance of ambiguities, this author proposes replacing the expression Closed Circulation of Energies (CCE) with *Voluntary Energetic Longitudinal Oscillation* (VELO), which will be used from now on in this article.

---

[1] A description of the OLVE technique can be found in Alegretti's and Sassoli de Bianchi's articles, published in this volume.

The VELO technique

As stated above, the profound knowledge and control of the *Voluntary Energetic Longitudinal Oscillation* technique, provides the consciousness with the ability to install a VS in *any* circumstance, according to the desired type and level of intensity.

In this technique individuals utilize their will and bioenergetic mastery to generate a longitudinal energetic pulse. However, the propagation of this pulse from one extremity to the other of the energetic body (which, in the majority of the cases, is coinciding with the soma) does not happen spontaneously, nor is it affected only by the normal 'resistance' of the energetic pathways.

Just like generating the energetic pulse, the maintenance of its propagation also has to be performed by the consciousness, as it requires employment of the same attributes and self-mastery applied in the initial production of this pulse. In this energetic maneuver, the pulse is synchronized via a specific type of paramotor coordination.

Therefore, without the consciousness' constant attention on and appropriate follow-up of the energetic pulse's flow throughout the energetic interface, such pulse normally loses coherence or dissipates, leading to energetic results other than the VS, or even producing no results at all.

The propagation of this bioenergetic pulse and the maintenance of the resulting oscillation depend on several mental and energetic attributes, which are the central focus of this work and are discussed in the section 'Basic Attributes for the Development of the VS' below.

2 Terminological clarification concerning the VS

Historical perspective and correction of the trajectory

Due to the natural human tendency of trying to look for more synthetic and simple ways of referring to a phenomenon, it became customary among members of the conscientiological community of researchers and students to refer to the complete process of the voluntary energetic longitudinal oscillation (or the closed circuit of energies) simply as 'VS.'

Individuals commonly say they will "do a VS" when actually they will make an attempt to mobilize their energy (VELO) aiming at installing the VS. Their session of bioenergetic exercise may or may not have a satisfactory result regarding the level and quality of the energetic mobilization, depending on the person's self-control over the attributes involved in the technique.

Although this condition is (or developed as being) only a 'way of saying it,' as time goes by it starts to foment basic misconceptions, mainly due to the fact that beginners in the energetic technique or students of conscientiology hear "to do a VS" instead of "to do the technique that has the objective of producing the VS."

It is important to highlight that when performing the technique to reach the VS, it is not implicit that the VS will definitely be reached. Hence, considering the result from the VELO technique could be or could not be the production of the VS, the question to ask when inquiring about the results of the practical energetic training – during a class, for example – should be, firstly, whether the person "was able to perform the VELO satisfactorily" and, secondly, "what was the result achieved". So, the question should not simply be whether the practitioner "installed the VS."

Asking students about results in the form of "how was your vibrational state?" predisposes the listeners to misconceptions about the VS and gives them the impression that reaching the VS at will is something trivial, superficial, and expected to be reached quickly and with ease.

Another area of error stems from this condition: taking personal results normally reached with the VELO – still imperfect – as being already the maximum level of personal energetic effect possible. This misconception favors a common self-corrupt pattern of speaking a lot about doing the VS without, however, applying the necessary effort to control the attributes of the VELO that will allow the installation and the complete mastery of the actual vibrational state.

It is not uncommon to find people who, in fact, have never reached a VS and do not understand what type of vibration or repercussion corresponds to this phenomenon. N.B.: *Not every vibration is a vibrational state.*

Terminology used in this article

In an attempt to look for clearer language to communicate the concepts presented in this study, the author 'borrows' basic expressions and concepts from other fields (conceptual migration), mainly from physics.

However, various attributes of the VS (and of the VELO) discussed here generate and suffer effects of multidimensional complexity. Thus, such expressions do not always have an exact conceptual equivalent to allow for a direct, precise and unequivocal terminological migration from physics to bioenergetics. Consequently, in this article, the use of some common expressions from the field of physics does not imply their linear and direct equivalence to the multidimensional-energetic context applied here.

Hence, all effort will be made in this work to provide commentaries and details about each attribute discussed, aiming to elucidate its exact context, concept, and definition.[2]

3 Methodology for measuring energetic attributes: history

Bases of experimentation and measurement

The experience from the bioenergetic and parapsychic evaluations of participants of the IAC's course called *Goal: Intrusionlessness*, developed and presented by Wagner Alegretti and the author, allowed for the devising, experimenting and testing of a methodology for measuring an individual's bioenergetic capacity.

The observations made through such bioenergetic-parapsychic evaluations resulted in the study called *Bases for the Energogram and*

---

[2] To favor the clarity and understandability of this study, the expressions that refer to attributes of performing the VS appear in italics. For example, the word 'depth' can appear without italics, having, therefore, its common meaning/significance; or it can be italicized, making clear to the reader that, beyond its common usage (acceptation), it remits to the concept of the attribute *depth* discussed in the article. Obviously, other expressions that grammatically or conceptually require italics (e.g., foreign expressions) will also be italicized.

Despertogram, which was presented in 2005 at the *Jornada de Despertologia* (Despertology Meeting) organized by the CHSC in the city of Foz do Iguaçu, Brazil. At this conference, the project plan, the basic parameters of this methodology, the practical aspects of bioenergetic measurement and the preliminary results of this study were presented to the participants.

The measuring scale of bioenergicity, developed by Alegretti and the author – which has been applied in private evaluation sessions of the *Goal: Intrusionlessness* course since 2003 – establishes a qualitative analysis and a quantitative gradation criterially arranged on a precise numerical scale. This scale serves to grade a broad range of diverse bioenergetic and parapsychic abilities, which are evaluated and worked on during such a course.

The abovementioned methodology to measure personal bioenergy, as well as the calibration of such a measuring system, are based on 1,084 hours of private sessions of bioenergetic evaluations and measurements of 294 students, performed as part of the *Goal: Intrusionlessness* course (as of October 2008).

The measuring of an individual's energetic condition and their capacity to control their own bioenergies is performed via the technical energetic coupling promoted by the researcher, who directs a series of energetic maneuvers that facilitates the testing and evaluation of individuals, according to a pre-established grading system.

Such a measurement and grading system – considering that it is based on the experience and comparison of results of more than 1000 session-hours of bioenergetic evaluations – offers the evaluated individuals a less subjective reference of their own condition.

Calibrating the measuring agent

Measuring someone else's energy requires, on the part of the measurer, a great deal of self-criticism, sufficient self-knowledge, and consistent energetic mastery. It is also necessary to have a clear *pre-established strategy*, based on solid protocols, so as to confer, from the beginning, uniform criteria on the observations, interpretations, and measurements recorded.

Despite the existence of such pre-established procedures, many of the strategies applied during the evaluation sessions mentioned above were implemented or perfected by direct suggestions from the team of nonphysical mentors who assist during the course.

Often these helpers brought the same inspiration to both instructors while conducting concomitant energetic evaluation sessions during the course *Goal: Intrusionlessness.* In other words, in those instances, such intuitions came simultaneously to both instructor-researchers during private evaluation sessions that were happening at the same time, yet, in different physical environments, where the instructors had no way of communicating with each other.

Such simultaneous inspirations, in many instances, worked as a confirmatory agent of the procedures applied in evaluation, analysis and bioenergetic training. Therefore, they also have a vital function for the instructor-researchers, since they work as an element of calibration and refinement of the measuring techniques and methods utilized.

As in any other measuring instrument, the bioenergetic evaluating agent (the researcher) has to maintain (1) his/her accuracy, through bioenergetic discernment, and (2) the smallest level of interference possible, through his/her cosmoethical self-scrutiny; reaching in this manner the maximum acuity and neutrality feasible during bioenergetic evaluations of others. The confirmations from intuitions, synchronicities, joint cognitions, as well as the coherent sensations and perceptions between the evaluator and the evaluated are another aspect taken into consideration when calibrating the researcher's measuring system. Considering that such evaluations have been performed since 2003, confirmations and inputs received *a posteriori* are an additional instrument for gauging and refining the measuring agents.

4 Basic attributes for the development of the VS

During the abovementioned private bioenergetic evaluation and measurement sessions, the author had the opportunity to observe directly some basic attributes involved in the VELO practice and, consequently, in the willful production of the VS through the mobilization of personal energies.

On these occasions, the author proceeded to register and catalog these attributes and, hence, the elements involved in the individual capacity of performing the VELO technique. The identification of such attributes occurred in a clear and unequivocal fashion, leading to the conclusion that it is possible to measure them through the bioenergetic coupling technique mentioned previously.

Among the aspects involved in the development of the VELO and in the control and installation of the VS, there are fundamental or primary attributes with direct implications and attributes that derive from the manifestation of other attributes. There are also attributes of an intraconsciential scope as well as compound attributes, where one is a variable of another, or where one element affects or interconnects with another, creating a stronger relationship between them or a binomial manifestation.

The measuring of students' capacity to control the VELO and the quality of their VS installation, performed during the private bioenergetic evaluation sessions mentioned above, allows for the examination of the quality of the application of all these attributes, regardless of their diverse categories and levels of importance.

It is worth mentioning that this study does not exhaust all the attributes and facets of the VELO or VS control. It merely discusses the attributes that could be better examined and deepened by this author to date. These attributes can be initially classified into:

1. Primary Energetic Attributes
2. Derived Energetic Attributes
3. Compound Energetic Attributes
4. Concurrent Intraconsciential Attributes

Real control of the VS – for the consciousness still in development regarding multidimensional self-awareness, energetic asepsis, thosenic health, quality of interconsciential relations, and qualification of its multiexistential holokarmic record – naturally demands that the consciousness "plows" through the VELO with honest and incorruptible self-effort. This path has to be covered without laziness, or devious excuses or shortcuts. That is:

The consciousness is not exempt from the VELO by having found a way of 'escaping' from it, but rather by 'facing' it and practicing it until reaching an absolute and permanent self-mastery over the VELO, hence, transcending it.

To perform the VELO efficiently (and, thus, to reach a broader bioenergetic self-mastery, which can make the intrusionlessness condition viable), the attributes described below have to be recognized, coordinated and mastered.

Primary Energetic Attributes

Primary Energetic Attributes refer to the essential or underlying attributes that form the VELO technique, which produce repercussions and ramifications that generate or permit the manifestation of other sets of attributes.

- *Primary attribute 1:* **Quantity**

Definition
   Quantity or percentage of consciential energy that the consciousness moves during the VELO.

Related concepts
   1. Quantity of energy mobilized or transported by the pulse.
   2. Equivalent to the pulse's amplitude or intensity.

Particularizations
   1. Given the natural total consciential energy of an individual's (CE_{TOT}) at a certain evolutionary level, according to his/her existential and evolutionary context, a certain percentage of this CE_{TOT} is, in general, more readily unimpeded[3] (CE_{FREE}), being, thus, the fraction of his/her energy that is possible for him/her to move with more ease. Note that the quantity of energy (Q) that the practitioner is able to move during his/her session of VELO varies from individual to individual and also from session to session, according to the practitioner's level of self-control. However, in practice, the initial magnitude of Q is, in general, approximately equal to one's CE_{FREE}, due to the fact that this is the fraction of one's CE that is spontaneously looser (i.e., at the beginning of the session, normally, $Q \leq CE_{FREE}$. Note that $CE_{FREE} < CE_{TOT}$).
   2. Through the determined application of his/her *will*, the practitioner can increase the amount of energy being moved (ideal condition) in a given session of VELO, which will lead to better results in that session. It is also worth mentioning that the continued execution of the VELO promotes the expansion of one's intrinsic percentage of CE_{FREE} (desired condition), leading to the

---

[3] According to the individual's pre-somatic condition (2nd desoma), inherent energetic looseness, and current circumstances.

amplification of general energetic health (energetic looseness).

Note

This attribute is directly related to one's capacity to perform a greater chakral or energetic unblocking.

- *Primary attribute 2:* **Fluidity**

Definition

Low holochakral 'impedance'. It is the opposite of energetic viscosity, resulting in the malleability or docility of the energy at the command of the consciousness.

Related concepts

1. Bioenergetic manageability.
2. Energetic looseness. .

Particularizations

1. Energetic *fluidity* is an intrinsic aspect of each individual, varying according to one's evolutionary level (and according to one's existential context).

2. In principle, the greater the level of *fluidity*, the greater one's CE$_{FREE}$ will be.

Note

The expansion of energetic self-control and, consequently, of the *quantity* of energy mobilized during the VELO, leads also to an increase in the degree of holochakral *fluidity*. In general, such increments occur throughout one successful session of VELO. However, with the accumulation of frequent well-performed VELO sessions, the natural (intrinsic) level of *fluidity* of the practitioner will increase, usually over months and/or years; in turn, making it easier for the individual to increase at will the *quantity* of mobilized energy in the VELO.

- *Primary attribute 3:* **Speed**

Definition

Property[4] inversely proportional to the time the energetic pulse takes to travel the energetic interface in a complete cycle (period). Average speed (scalar speed) of the pulse while travelling the

---

[4] For disambiguation, 'property' instead of 'quantity' is used here, so as to avoid confusion with the energetic attribute here called *quantity*. In mathematics/physics 'quantity' is the fundamental term used to refer to any type of quantitative/<u>measurable property</u> or attribute of things.

entire cycle coronochakra-solechakras-coronochakra. N.B.: the instantaneous speed is zero in the extremities, immediately before the sense[5] is reversed.

Related concepts

1. Frequency of the energetic pulse.

2. Length of time the pulse takes to travel the energetic interface from one extreme to the other.

3. Scalar speed of the energetic pulse.

Particularizations

1. In a complete analysis of a specific VELO session, the frequency of the oscillatory movement can be considered (the frequency of the energetic oscillations is greater when there is greater speed of the longitudinal movement).

2. A central point in the VELO procedure is to increase the frequency throughout the execution of each session.

Notes

1. The average (scalar) speed studied here is directly proportional to the frequency, the latter being a more suitable property to express this parameter. Therefore, in favor of technical precision and accuracy, the term frequency should be used instead of speed. However, since the term frequency is generally harder to understand for the ordinary practitioner, 'speed' (S) was the word chosen to express this attribute, opting, in this way, to utilize a simpler term that allows a more intuitive comprehension of the concept.

2. The presence of a specific chakral blockage can cause a reduction in the speed of the pulse in the region corresponding to that chakra. Usually, when moving energy in parts of the body other than the referred region (i.e., after passing through the energetically blocked area), the practitioner recovers his/her average *speed*.

- *Primary attribute 4:* **Sweep**

Definition

Interval of space covered by the longitudinal oscillatory energetic pulse.

Related concepts

1. Length of the path of the energetic pulse.

---

[5] *Sense* as in vectors (mathematics and physics), meaning *orientation along a given direction*; i.e., in a vertical "direction" there are two "senses": up and down.

2. Spatial amplitude of the propagation of the bioenergetic pulse.

Particularization

Total or partial coverage of the energetic interface in the energetic flow.

Note

In the VELO, the energetic sweep must cover the total extension of the energetic interface, i.e., from the top of the parahead to the soles of the parafeet.

- *Primary attribute 5:* **Rectilinearity**

Definition

Quality of the energetic flow in a straight and direct line through the energetic interface, without spirals, curves, circles, sinuosities, detours, or unnecessary movements that corrupt the rectilinear sweep of the longitudinal oscillatory energetic pulse.

Related concept

Straightness of the route of the VELO.

Particularizations

1. Preservation of the ideal route of the pulse's *sweep* during the VELO.

2. The rectilinear flow favors the sweep of the entire holochakra with the greatest possible efficiency, as it establishes the shortest route – straight and direct – passing through the energetic interface from end to end, allowing the completion of this trajectory with the least 'expenditure of energy' in the performance of the VELO per se.

Notes

1. Due to inexperience or simply lack of energetic control, the practitioner commonly allows the emergence of curves or detours in the movement of energy during the VELO. At other times, in search – inappropriately and mistakenly – for alternative ways to move energy, the practitioner consciously or unconsciously generates an energetic flow in a spiraling or other non-rectilinear form.

2. Many times, the existence of energetic blocks is what generates the detours in the energetic flow. In this case, instead of continuing in a laminar (straight) flow through the energetic interface, the energy establishes a type of 'energetic turbulence', which affects its *rectilinearity*. When the movement of energy spontaneously

curves or detours, it is because it is deviating from certain chakras or regions; therefore, logically, it is not reaching all parts of the energetic interface in the most direct and efficient manner possible. In other words, without enforcing the *rectilinearity* of the movement, the energetic blocks may remain untouched.

3. The sinuosities and turbulences in the energetic flow make it harder to reach energetic resonance, since they affect the coherence of the energetic regimen that leads to the VS.

4. If the practitioner lacks the coordination to move energy in a straight line (simpler route), it is unlikely that he/she will be able to coordinate more complex movements such as, for example, a spiral form, with enough dexterity to reach the highest level of excellence possible regarding *quantity*, *depth*, and the other attributes studied here. N.B.: When practitioners find that it is easier to move their energies in a non-rectilinear fashion, this is usually due to the fact that, in such energetic maneuvers, they end up moving only the looser (generally, more superficial) energies of their energetic interface. This leads them to achieve more immediate and easily identifiable sensations, which are, nonetheless, lighter and more ephemeral. In general, such sensations do not correspond to the *activation* of the energetic interface and this procedure (of superficial mobilization of energies) does not deliver the complete benefits of the VELO, as, in this case, the diverse attributes required in the control of the VS studied here are not manifested.

Derived Energetic Attributes

These attributes refer to the VELO elements which are possible to manifest from the existence and expression of other attributes. The derived attributes mentioned below arise from the primary attributes and are formed due to different interrelations among themselves or with other parameters.

● *Derived attribute 1:* **Consistency**

Definition
VELO without any reduction or undesirable alteration in the amount of energy moved; thus, keeping the VELO's *quantity* unaltered or implementing the appropriate increase in *quantity*.
Related concepts
1. Linearity (linear variation) of the *quantity*.

2. Absence of unnecessary fluctuations in the *quantity* of energies moved.

3. Regular progression of the *quantity*.

Note

In the VELO, the ideal condition is to institute a stable and linear increase of the *quantity*, establishing a sustained/steady progression throughout the session.

• *Derived attribute 2:* **Rhythm**

Definition

Continuity and stability of the speed of propagation of the pulse and of its subsequent acceleration (including the time period taken to reverse its sense).

Related concepts

1. Maintenance or linear progression of the speed of the oscillatory movement.

2. Cadence of the energetic pulse throughout the VELO.

3. Stability of the speed and, subsequently, of the acceleration of the energetic pulse.

4. Level of regularity of the acceleration.

Particularizations

1. VELO without interruption, abrupt changes, or inappropriate fluctuations in the *speed* or frequency; i.e., there is no 'little stop', rest, or pause to regain concentration or to observe the sensations.

2. Analogous to tempo, in music.

Notes

1. Linearity of speed (in function of time, from path to path) is a fundamental factor.

2. The acceleration of the energetic pulse is supposed to occur in a continuous fashion, without any abrupt alteration in its *rhythm*. Therefore, it is expected that the *rhythm* will not be constant during the entire VELO; however, it is the smooth progression, with a gradual, even and linear increase in speed that is the proper condition for the VELO and favors the installation of the VS.

3. At times, beginners allow an inappropriate fluctuation of speed in the energetic pulse from one path (half cycle) to another. They may also apply different speeds within the same path. In both instances, they fail to apply the *rhythm* to the VELO.

4. Ideally, to reach the maximum average *speed* in a given path,

beginners should rapidly accelerate the energy, right after reversing the sense of the pulse, maintaining the speed at its maximum during the entire path. They must decelerate it only when very close to the end of each path, i.e., just before proceeding to invert the pulse in the opposite direction.

• *Derived attribute 3:* **Depth**

Definition
Complete penetration and action/effect of the energetic pulse through each segment of the body, inasmuch as the pulse deepens spatially through the energetic interface and also multi dimensionally, i.e., reaching beyond the more 'accessible' layers of the energetic interface and, as a result, reaching 'more rigid' (blocked, crystallized, old, pathological, deep, or fossilized) thosenes.

Related concepts
1. Breadth and reach of the energetic pulse.
2. Excellence in the energetic sweep of the energetic interface.

Particularization
The VELO can, at one extreme, affect and move only the looser, more manageable energies (CE$_{FREE}$); or, at the other extreme, include the 'core' of the energetic interface, promoting a deep energetic shake-up and, therefore, reaching and affecting the stagnant and evolution-hindering energies, including retrothosenes and retropsychic scars.

Note
Energetic blocks – either generalized (low *fluidity*) or of specific chakras – diminish the level of *depth*, creating, in the process, a vicious circle that has to be broken by the individual via the optimal application of the *quantity-depth binomial*.

• *Derived attribute 4:* **Compaction**

Definition
Degree of spatial concentration of the energy in the pulse.

Related concepts
1. Quality of the propagation of the bioenergetic pulse.
2. Compactness of the energetic propagation.
3. Pulse width.

Particularization
For example, there is: (1) definite propagation, which has neither

dispersion nor reverberation that would dissipate the strength of the energetic pulse; or (2) disperse propagation, with the formation of an energetic 'trail' of the pulse.

Note

If the total *quantity* of energy moved (Q) is equal to the quantity of energy that passes, at each half-cycle (one path), through a given transversal section of the soma, *type 1* of the above-mentioned propagation occurs (ideal condition). If, however, the energy that passes through a given transversal section of the energetic interface is lesser than Q because part of it is at that moment still passing through the prior transversal sections of the energetic interface due to a pulse delay or drag (dispersion of the pulse intensity), then, a *type 2* propagation occurs (undesirable condition).

- *Derived attribute 5:* **Activation**[6]

Definition

Condition of intensification of the energetic power, resulting in the general or partial chakral-bioenergetic activation of the energetic interface.

Related concepts

1. Partial or generalized bioenergetic intensification reached.
2. Energetic dynamization or activation.

Particularizations

1. Result of the synergistic combination of the attributes of the VELO, being an attribute directly correlated with the excellence in the application of the primary attributes.
2. When the *activation* reaches a certain level that produces <u>bioenergetic</u> <u>resonance</u> in the entire energetic interface, it is classified as a VS. The VS is, therefore, proportional to the magnitude of the energetic *activation* (A) or of the attained resonance. Hence, different degrees of resonance will produce different levels of VS intensity, with different effects and repercussions.

Notes

1. The *activation* can occur in one chakra or in a few chakras, or can affect the entire energetic interface, the latter being the ideal

---

[6] See in the section "The Energetic *Activation*" below, pertinent notes about this attribute, which, for practical purposes and for measuring the vibrational state, is the VS per se.

condition sought, which, depending on the intensity, can be considered a vibrational state.

2. Not all *activation* corresponds to a vibrational state, since it does not always reach a complete resonance of the energetic interface or is not always self-sustained adequately enough to a point it could be considered a VS. A minimum level of energetic activation (A_{MIN}) is needed for it to be considered a VS, regardless of the intensity that this VS (I_{VS}) will have, i.e., the VS occurs when $A \geq A_{MIN}$. N.B.: Is still unknown how to measure A_{MIN}.

3. In rare cases, the resonance reaches more vehicles of manifestation, beyond the energetic interface, including possibly even encompassing all bodies. N.B.: It is more probable for such a condition to occur when the intraphysical consciousness is in a state of deep relaxation or if the bodies are partially non-aligned.

Compound Energetic Attributes

Compound attributes of the VELO refer to the existing inter-relation between the primary and the derived energetic attributes, or between energetic attributes and other factors[7] which are also involved in exercising the bioenergetic self-control necessary to perform the VELO.

The inter-relation between the components of a compound attribute can have inherently different qualities, thus, leading to associations, interferences or synergies among the elements that constitute such an attribute. It can, therefore, affect the result of the technique as well as the consciousness' capacity to produce a VS at will, that is, even when under external opposing pressure, with personal difficulties, or with inner afflictions or conflicts.

The referred link of inter-relation can be due to (possible) interference that one component of the attribute may generate over the other, so as to affect, nullify, or corrupt it, forming a relation of affectability in its manifestation. Such a link existing between the elements that forms the compound attributes can also be of a synergistic, complementary or intersectional character, forming a type of manifested binomial.

---

[7] E.g., consciential attributes.

- *Compound attribute 1:* **Quantity–Speed relation**

Definition

Capacity to maintain the *quantity* of energy moved stable or expanding while willfully increasing the *speed* of propagation of the energetic pulse.

Related concept

Invariance (or increment, if applicable) in the *quantity*, despite the increase in *speed*.

Note

Reducing the *quantity* when increasing the speed of the VELO is a common condition for beginners.

- *Compound attribute 2:* **Depth–Speed relation**

Definition

Maintaining a high degree of comprehensiveness and reach of the energetic pulse (*depth*), independently of the applied *speed*.

Related concept

Complete penetration and permeability of the energetic pulse through the interchakral channels, even when promoting an increase in the VELO's *speed*.

Note

The inexperienced practitioner usually ends up reducing the *depth* of the energetic flow when promoting an increase in the frequency (i.e., accelerating of the pulse).

- *Compound attribute 3:* **Sweep–Speed relation**

Definition

Maintaining the propagation of the longitudinal energetic pulse through the complete path (from the top of the parahead to the soles of the parafeet), regardless of the applied *speed*.

Related concept

Totality of the energetic *sweep* throughout the energetic interface even with the increase in *speed* of the VELO.

Note

In general, the beginning practitioner has the tendency to reverse the sense of the pulse before reaching the end of the path (coronochakra or solechakra) when promoting the increase in speed.

- *Compound attribute 4:* **Rhythm–Speed relation**

Definition
  Maintaining the *rhythm* or oscillation of the pulse, independently
  of the *speed* applied.
Related concept
  Level of regularity of the oscillation of the energetic pulse during
  the development of the entire session, even while increasing the
  VELO's *speed*.
Note
  Usually beginners fail to maintain the rhythm, or lose control of
  the energetic movement, while trying to accelerate the pulse. This
  leads them, many times, to limit the speed as a way of guaranteeing
  the quality of the *rhythm*.

- *Compound attribute 5:* **'Sustaining the oscillation'–'Application
of effort' relation**

Definition
  Effectively maintaining a level of excellence of the VELO, no
  matter what amount of effort is required.
Related concepts
  1. Unwavering and uninterrupted control of the quality of manifestation of the primary attributes of the VELO.
  2. Maintaining the effort versus maintaining the result.
Notes
  In general, the consciousness establishes his/her reference point
  on the amount of personal effort applied, thus, keeping this element fixed. Consequently, when energetic blocks or interferences
  arise, the oscillatory movement ends up getting reduced due to the
  decrease in *fluidity*, which makes the mobilization of bioenergies
  more difficult. N.B.: the correct condition is the unwavering and
  steadfast maintenance of the VELO, taking into account all the
  attributes and guaranteeing the desired progression of the *quantity*
  and *speed*, regardless of the obstacles or difficulties that may appear
  (i.e., maintaining the level of excellence in the result obtained).

- *Compound attribute 6:* **Rectilinearity–Depth binomial**

Definition
  Direct effect that *rectilinearity* can have on the optimal application

of the VELO's *depth*.

Related concept

Superficiality of the energetic flow or reduction in the *depth* of the penetration of the energy in the entire energetic interface's interchakral channels. Such superficiality is caused by the detours, curves, or turbulences formed during the process of mobilization of energy throughout the energetic interface.

Notes

1. *Rectilinearity*, associated with suitable *quantity*, guarantees that the energy will pass through all points of the energetic interface (completeness); thus, it has a direct relation to the level of *depth* the energy will reach in the VELO. N.B.: If the energy deviates, presumably, it is detouring from certain areas where there are probably energetic crystallizations or blocks connected to retrotraumas, revealing, therefore, that the energy is flowing inefficiently through the energetic chakral pathways.

2. If, on the one hand, the *rectilinearity* can help in obtaining greater *depth*, the opposite is also true because, given that the application of optimal *depth* favors energetic unblocking, the rectilinear flow of energy will be facilitated upon reaching greater proficiency when employing *depth*.

• *Compound attribute 7*: **Quantity–Depth binomial**

Definition

The effect that the *quantity* has over the *depth* of the energetic flow and, as a result, on the unblocking power of the VELO.

Related concept

Level of completeness, penetration, and spread of the energetic pulse through the energetic and interchakral channels (i.e., *depth*), as a consequence of the proper application of the *quantity*. N.B.: Such a condition can promote the amplification of the amount of CE_{FREE}.

Note

The greater the energetic *quantity* and intensity of the pulse, the greater is the possibility that it will flow through all of the chakral interconnecting circuits, reaching greater *depth* and being able in this manner to promote even multivehicular (in more than one body) repercussions.

• *Compound attribute 8:* **Action–Relaxation binomial**

Definition
An apparently paradoxical synergistic combination applied by the practitioner of: (1) the ability to apply direct control over one's bio-oenergies (active will; energetic action commanded by the mental body) with (2) inner relaxation (i.e., lessening of expectations or anxiety) to allow the rise of the holochakral dynamization (the VS).

Related concepts
1. Self-control – Acquiescence Binomial.
2. Willful non-somatic action.

Notes
1. Many times the VS *comes to the practitioner* as a result of energetic looseness and activation obtained through his/her active mobilization of energy.
2. In certain cases the VS is facilitated or intensified by helpers during the VELO with diverse objectives, including assisting in the overall unblocking of the practitioner.
3. Without the correct posture of inner openness, the practitioner may, inadvertently, slow down the occurrence of the VS.

Concurrent Intraconsciential Attributes

Some consciential attributes are more closely connected to the result and quality of the VELO; therefore, their manifestation directly affects one's bioenergetic self-control and the efficient mobilization of energies.

In fact, it can be affirmed that if an intraphysical consciousness fully applies the consciential attributes listed below in his/her exercise of energetic mobilization, on some level and to a certain degree, the VELO will occur. This leads to the gradual improvement of the individual's energetic condition and bioenergetic control and, hence, in time, to the VS. N.B.: Such a condition is true even if the practitioner does not feel his/her own energies during the execution of the VELO.

• *Consciential attribute 1:* **Intent**

Definition
Intellectual-mental decision, arising from a deep inner understanding of the value of a certain objective, leading the

consciousness to the intention and legitimate decision of looking for ways of reaching it.

In this context, it refers to the act of really wanting, *at any cost*, to perform the mobilization of the personal CEs, regardless of the existence or probable emergence of difficulties or obstacles.

Related concepts

1. Legitimate inner decision.
2. Personal resolution.
3. Discerning and contextualized interest.
4. Firm personal choice.

- *Consciential attribute 2:* **Will**

Definition

Inner determination that impels the consciousness to carry out his/her established objective.

Related concepts

1. Internally guided effort.
2. Persistence of action/execution.
3. Underlying element in self-control.

Particularization

Will refers to the quality of effort and the applied inner diligence; being, in this case, the attribute which makes the crowning of the *intent*. It brings about successful energetic work by means of applying the effort necessary for the efficient execution of the VELO. Being the materialization of the *intent*, *will* is a key element in generating the energetic pulse and, consequently, promoting the VS.

Notes

1. Main element, generator or maintainer of the non-dissipation of effort, which assures the focus of one's action/activity going directly and exclusively to the energetic interface.
2. *Will* is the factor responsible for the application of indefatigable personal dedication over time, until the desired objective is reached.
3. Underlying resource for the best possible manifestation of *sustaining the oscillation – application of effort relation*.
4. Indispensable factor for indisputable self-control.

- *Consciential attribute 3:* **Attention**

Definition

Capacity to maintain focus during the complete execution of the technique, without mental interruptions, distractions or day-dreams.

Related concepts

1. Concentration.
2. Non-dissipation / no distraction.

Notes

1. *Attention* is the foundation which allows the very execution of the VELO, since without guided and superior focused attention the consummation of the longitudinal energetic mobilization is compromised, even if the practitioner has, as an intrinsic potential, the capacity to control the VELO attributes.

2. It is common for the practitioner to become distracted with external stimuli, somatic sensations, or even energetic-chakral sensations generated by the exercise.

3. Spontaneous thoughts which occur during the VELO may dissipate concentration, mainly if they are connected to emotions or are incited by thosenic intrusion.

5 The installation of the VS

Technique variations

As observed, the procedure of the VELO technique is extremely simple. Yet, to attain maximum results (according to the potential of each consciousness), attention to the quality of the application of the attributes involved is necessary.

Even when practitioners invest time and effort and take care in the application of each attribute of the VELO, natural variations in its execution may occur, according to the style and predisposition of each individual. Note that, in order for them to be in fact variations only (still maintaining the objectives, effects and benefits of the VELO), they cannot dispense with the basis of the execution of the technique.

For example,[8] it will make no difference in the result if the practitioner applies one of the following variations.

1. Start the energetic longitudinal mobilization from the coronochakra or from the solechakras. In any of these conditions, once each of the subsequent energetic paths has a complete *sweep*, the result will be the same.

2. If the VELO is started by a specific chakra – different from the ones mentioned in item 1 above – the result of the technique will also not be compromised as long as, after the energy mobilization has started, the pulse continues with a constant and complete *sweep*. N.B.: In order to better sense and command the energy, some people prefer to start the VELO from the chakra they have a greater predisposition to feel.

3. Focus more intensely in the inversion of the pulse's sense (meaning, when the energy reaches one of the energetic interface extremities; or when the energy starts to go up if it was going down, and vice-versa), by making a stronger push at this point. This resource, in general, serves for some as a way of maintaining the pulse's quality (for example, of its *rhythm* and *consistency*).

Other variations acceptable for practitioners who are still beginners in the execution of the technique or in controlling the VELO attributes are:

1. When a dispersive and delayed propagation of the energetic pulse takes place, some individuals continue pushing the energy to the extremity of the energetic interface; thus, only reversing the sense of the energetic pulse when the total *quantity* (Q) of the energy moved along that path has reached the target-extremity of the energetic interface (i.e., the coronochakra or solechakras). For didactic purposes, it can be said that it is as if the intraphysical consciousness 'waited' for the delayed energies to arrive at its final destination before changing the sense of the energetic pulse. N.B.: In certain cases this procedure is beneficial, since it avoids the practitioner moving an even lesser quantity of energy than his/her CE_{FREE} due to

---

[8] The small list below is just an illustration to provide the reader with examples of the type of variations that normally can occur without negatively compromising the VELO result and without corrupting its correct execution.

experiencing the 'propagation *type 2*' mentioned in the item 'compaction' of the derived attributes. However, the condition that leads practitioners to manifest this type of energetic propagation (non-ideal) has to be identified and overcome as soon as possible, so that they establish a cohesive and appropriate propagation of their energies.

2. Some people opt to do, for a brief moment, an energetic mobilization in smaller segments of the energetic interface (for example, from the frontochakra to the umbilicochakra) before performing the VELO technique, that is, before making the pulse have a complete *sweep*, traveling the entire energetic interface from one end to the other. In some cases, such a procedure helps in maintaining focus (*attention*) and in the initial or partial unblocking. N.B.: It is important to highlight that such modular energetic mobilization (in modules of energetic interface segments) does not replace and does not have the same effect as a VELO; hence, it cannot be taken for the VELO technique. It is only a pre-VELO procedure to 'warm-up,' which should eventually be abandoned.

The Time Factor

Even though *time* (duration of session) is a variable of the execution of the VELO, it is a factor extrinsic to the consciousness; therefore, it differs from the above discussed attributes in its application, role and influence on the VS.

For beginners, the *time* (T)[9] that they apply in executing the technique is, in general, important. Such is the case because, considering their level of *fluidity* and overall self-control, the higher the number of energetic cycles they perform in a specific session, the greater will be their chances of succeeding, as they will have better conditions to potentialize their energy and arrive at a state of resonance of the energetic interface.

A VELO session is an <u>uninterrupted</u> mobilization of personal CE (consciential energy, bioenergy). Thus, if a practitioner does 40 minutes of VELO, but becomes distracted approximately every 2 minutes – therefore interrupting or affecting the session – he/she

---

[9] The capital letter "T" is used here to refer to 'time' meaning duration of the session.

would have done 20 mini-sessions and not one long session of 40 minutes during this time.

The execution of one long session produces cumulative effects that bring certain benefits as well as palpable positive impacts, clearly identifiable, to the practitioner's set of bodies. Such results, however, are unlikely to be obtained at the same level with the execution of many consecutive mini-sessions as described in the previous paragraph.

Besides the obvious fact that the accumulation of effects does not happen in the same way with many consecutive mini-sessions as with one long (uninterrupted) session, the occurrence of a series of non-planned interruptions or pauses also demonstrates a lack of energetic self-control of the practitioner and, consequently, the absence of a suitable application of the attributes discussed in this article. Such an occurrence reveals, therefore, that the practitioner is not performing the VELO correctly or according to the maximum of his/her intrinsic potential.

When the individual still needs to acquire greater self-control over the VELO's (and the VS's) attributes, the form that brings most results is performing many sessions of the technique distributed throughout the daily waking state period. As a general rule, it is suggested that he/she performs, for a certain period of time[10], 20 daily sessions spread out preferably in equal time intervals. N.B.: The practitioner, or his/her trainer/evaluator, may come to the conclusion that fewer daily sessions are enough or, conversely, that the practitioner needs to perform more than 20 sessions a day to be able to produce the desired results.

Even if the practitioner is not able to reach the VS (due either to the *fluidity* not being ideal or to other attributes not being well developed), for the purpose of training and acquiring control over the VELO attributes, doing sessions of approximately 5 minutes each – uninterrupted and without distractions or self-corruptions – should produce an improvement in his/her control over these attributes.

Even though the VELO is a simple procedure, in the majority of cases, months or even years of dedication are generally necessary before the average practitioner is able to reach a level of *fluidity* and parapsychomotricity (Alegretti, 1992) that would allow him/her to

---

[10] Days, weeks, or even several/many months in some cases.

improve the application of *quantity* in his/her exercise. This would, in turn, also lead the practitioner to a greater control over the diverse VELO and VS attributes as well as to a more effective and permanent expansion of his/her CE_{FREE}.

During this time, the practitioner will perceive a progression (usually non-linear) of his/her self-control; since, by correctly performing the VELO, greater results will be reached in shorter sessions.

It is important to clarify that if the practitioner executes only the *tensional state*,[11] not even 60 daily sessions of 5 continuous minutes or 5 complete, uninterrupted sessions of one (1) hour, performed throughout the day, will bring results.

It can be inferred from this that even though (depending on the practitioner's personal condition) sustaining the VELO for a considerable period of *time* can be critical for reaching a reasonable level of energetic *activation*, the importance of the duration of the exercise is interconnected with the quality of the manifestation of the VELO attributes.

Thus, the relevance of the *time* factor, in comparison with the other attributes, is in general inversely correlated to the control of the attributes discussed here. That is, without a certain minimum level of control over the VELO attributes, the probability of installing the VS is smaller, even with the execution of the technique throughout a sufficiently long period of *time*. On the other hand, the better the application of the VELO attributes, the less time the individual will need to reach results.

Nevertheless, despite this apparent paradox, it is the practice of the VELO itself that will lead the individual to achieve control over these attributes (even if the VS is not reached or if the energies are not clearly perceived). Therefore, it is essential to invest care and effort to reach and maintain excellence in the VELO. N.B.: When the intraphysical consciousness achieves greater actual control over the VELO attributes, frequently, a high-level VS can come about within a few seconds of energetic mobilization; meaning, with a VELO of only one or a few cycles.

---

[11] See the section "Facets of the VS."

The Energetic Activation[12]

The intensity of the VS (I_{VS}), derived from the quality of the energetic resonance, depends directly on the *quantity* (Q) of energy mobilized in a given period of time. Hence, the *speed* of the cyclic longitudinal pulse (S) is relevant, since, if the *speed* is higher,[13] the number of times that particular quantity of energy will pass through each specific transversal section of the energetic interface per unit of time will be greater; therefore increasing (from the point of view of the energy that flowed through the interchakral channels) the effect of the energetic mobilization.

The above-mentioned speed (S) can alternatively be defined as the number of times or cycles (N) that the specific *quantity* of energy passes through the energetic interface divided by the period (T) of the session ($S = N/T$). If we assume a given case in which Q is maintained constant during the VELO and if we establish a parallelism between *activation* (A) and the physical concept of power (i.e., the quantity of energy that passes, on average, through the energetic interface per unit of time), it can be affirmed, in a simplified way and for pedagogic purposes, that $A \sim Q \times N/T$ or $A \sim Q \times S$. Meaning, the *activation* is proportional to the *quantity* (Q) multiplied by the *speed* (S).

The *rhythm* of the VELO is essential, as it will give regularity to the pulse, a condition which will contribute towards reaching a frequency favorable to the energetic resonance or *activation* to such a level that it can be considered as installing the VS (A_{MIN}).

Note that there are vibrational states that occur in a spontaneous fashion (or, more accurately, apparently spontaneous), which are sponsored by helpers or brought about by other physical, nonphysical, mental, or bioenergetic factors. Therefore, in these cases, even if the Q or S produced directly by the practitioner could be equal to 0 (zero), the VS could occur.

In the case where the VS is promoted through the VELO, it is

---

[12] The formulas here presented aim solely to express the information in a 'different way,' which, depending on the reader's academic background or intellectual style, might contribute to the comprehension of the topic. Thus, for readers who find difficult to read formulas, the information can be captured through the conceptual and descriptive discussions of the section.

[13] As long as the mobilized *quantity* is not reduced (*quantity–speed relation*).

worth mentioning that without the mobilization of a considerable *quantity* of energy, it will be difficult to produce a resonance of an appropriate level no matter the duration of the exercise (T), because the *activation* will probably still be insufficient.

Thus, as a conjecture, I propose here a phenomenological formula more complete than the prior one (however, still only indicative of the possible relations of some of these attributes with the energetic *activation*). Such a formula, which aims at expressing the installation of the VS, includes the *time* factor, as follows: $A \sim Q^a \times S^b \times T^c$, where a, b and c are factors or weights – still unknown – that, in this case, are dimensionless positive numbers greater than zero which establish the relative or proportional contribution of each of these factors to the *activation*.

According to the experience of this author, it seems that a > b > c. In other words, the *quantity* is more relevant than the *speed*, while the duration of the session is the least relevant factor (i.e., the correct application of the discussed attributes in this article can be more relevant than the duration of the VELO).

Certainly, there are other factors that interfere in the installation of the VS and could be inserted in this formula; however, we are still far from knowing the quantitative and qualitative relation of them to the VS.

It is also important to highlight that the above formula seeks to translate qualitative information; that is, it merely illustrates, for didactic purposes, the aspects discussed in this section, called 'The Installation of the Vibrational State.'

Comparing various sessions of the same practitioner, it was possible to perceive that each of the VELOs performed were often different. They differed not only in their execution but also in the sensations and effects they generated. The same applies to the VSs reached. This fact increases even more the complexity of establishing standardized patterns for such phenomenon.

It is worth remembering that the *activation*, or chakral resonance, is affected also by the compound, derived, and intraconsciential attributes, which cause the nullification, reduction, or potentiation of the results; hence their importance for the installation of the VS at will and the complexity of arriving at a standardized and precise form to express the phenomenon.

Facets of the VS

As can be inferred from the discussions so far (and was previously mentioned), often there are situations in which – due to the concurrence of external variables or lack of personal bioenergetic self-control – beginners are not able to install the vibrational state during their sessions of energetic exercise. Nevertheless, it is worth re-iterating that in the great majority of cases only by doing the VELO (even if not reaching the *activation* of the bioenergies or the VS) can one observe tangible benefits of diverse levels, types, ramifications and repercussions. Therefore, the more meticulous attention and dedication the intraphysical consciousness applies to correctly executing the VELO, the greater the positive results will be.

Generally, less experienced practitioners have difficulty discerning what they experience, not being able to distinguish between (1) the energetic sensations produced only by the mobilization of energy during the VELO, (2) the energetic results obtained as a result of the correct and diligent performance of the VELO, and (3) the energetic resonance (A) or the VS installation *per se*.

There are vibrational states and **Vibrational States**! Not all VSs are equal; so, once a VS is installed, its intensity (I_{VS}) and nature are still to be verified. Different VSs can produce different repercussions and effects on the practitioner.

It is also worth clarifying that, once the vibrational state is achieved, i.e., the specific level of resonance of the entire energetic interface is reached, the manifestation of such a resonance superimposes itself on the other attributes.

Among the frequent misconceptions in the appreciation of the personal results obtained is the fact that (inexperienced) practitioners may mistake the energetic interface for the soma. That is, they regard the effect of somatic tension (inappropriate) – and resulting tremor or physical movement – as the energetic interface's *activation*. In this case, these practitioners promote basically somatic sensations that, in general, tire them and may even affect their cardiac frequency, blood pressure, and muscle tonus, effects that are erroneously taken for the 'vibratory activation' of the VS. This condition, which resembles more a 'trembling state,' is surprisingly common.

AutoRicerca, Issue 20, Year 2020, Pages 51-87

6 VS: defining its quality

An entire, separate article is needed to present a deeper analysis and understanding of the nature, intensity and quality of the VS.

Nonetheless, for the purpose of providing a broader context and a greater grasp of the attributes involved in the self-control of the vibrational state – the central objective of this work – it is prudent to offer here a few general comments about this topic.

The nature of the VS, or chakral *activation*, in a basic and initial analysis, can be roughly classified into 2 fundamental types:

- **Superficial activation**

Partial activation, characterized by: 'shallow', unstable, incomplete, ephemeral vibrations. Insufficient energetic stimulation, which is, in the majority of cases, restricted to the superficial, more external layers of the energetic body.

- **Profound activation**

Total activation, characterized by: complete, self-sustained, more stable and *deep* vibration with greater 'wave amplitude'.

Energetic dynamization that reaches a higher percentage of the energetic interface (and sometimes even of other bodies of manifestation), being more lasting than *'partial activation'*, maintaining more cohesion and coherence when installed.

The consciousness can also experience, in principle, 2 basic types of VSs, according to the degree of energetic intensification or repercussion produced (which may or may not be consistent with his/her intrinsic potential):

- **Comforting**

When the consciousness moves only his/her already flexible, manageable and partially activated energies, acknowledging the mobilization of such energies as relatively easy or more within his/her normal capacity.

Note: In this case (due to inexperience, lack of skill, or self-corruption) individuals usually end up mobilizing a small percentage of their total *quantity* of energy, making an erroneous assessment of the quality of their VELO, thus, being satisfied with an unsuitable or insufficient result.

Thus, commonly, the intraphysical consciousness continues to move a minute quantity of energies, in general reaching only the 'superficiality' of his/her energetic interface. This situation usually occurs also because the practitioner seeks, internally, the consoling or comforting notion that he/she is performing the technique correctly and reaching the highest level of VELO. In this fashion, the individual avoids critical self-evaluation and the arduous work sometimes required for a satisfactory realization of the VELO.

- **Stirring-up**

When the intraphysical consciousness is able to reach a level of *depth* and intensity that goes beyond his/her average level, the VELO 'touches'/affects the inter-vehicular[14] energetic connections where consciential energies with old patterns, undesirable for evolution, reside.

Given this fact, a VELO executed at a high level in terms of its *quantity* and *depth* can lead to the exposure of existing energetic blocks, some of them, from many lives.

This type of chakral intensification (I_{VS}) is, therefore, more self-antiseptic, produces more renovation of the energies, and promotes a better intraconsciential recycling, leading to an improvement in *fluidity*.

Note 1: 'Shaking up' the energies that were stagnated and repressed in the chakral and intraconsciential depths commonly produces an uncomfortable sensation, which can be quite ephemeral if the consciousness knows how to move forward with the process until the unblocking is achieved and the self-intoxicating energies are purified or removed.

Note 2: Although, due to the sensation it produces, this repercussion can be incorrectly perceived as negative, the consequences generated by it are positive and reveal a high level of *depth* reached with the energy work.

Unfortunately, in the majority of cases, practitioners apply various

---

[14] Inter-bodies.

84

self-corruptions and end up reducing or interrupting the mobilization of their energies when they feel such repercussions, so as to escape from the inner shake-up and the self-confrontation that can be brought about by this condition.[15] N.B.: Just like, for example, a physiotherapy session can generate discomfort (and, in some cases, even pain) which is fleeting and expected in the therapeutic process, tackling certain stagnated energies can also produce uncomfortable sensations when the unblocking (positive result) is taking place.

Note 3: The repercussions of stirring up blocked energies – previously hidden – normally occurs when the consciousness reaches a certain minimum level of *depth* in the VELO (in that specific session or in a series of sessions that he/she had been doing).

Note 4: It is worth stressing that this condition is different from the discomfort or negative pressure practitioners suffer (when trying to execute a VELO session) due to anticosmoethical interference from less lucid nonphysical consciousnesses who seek to impede them installing the VS. This subterfuge is used by those nonphysical consciousnesses as a way to hinder the practitioner's development, which, in turn, affects the practitioner's capacity to become free of their corrupt influences.

7 Future research

The research findings, theories, propositions and discussions presented in this paper can be refined by a longitudinal study of diverse methodology.

A detailed compilation of grades and results of each VELO execution and each attempt at installing a VS performed by a group of practitioners being studied can allow the identification of (1) the elements statistically easier to control, as well as (2) the more complex aspects of the VELO.

Such a compilation must be carried out through lucid and accurate self-measurement (by the practitioner) and also through hetero-

---

[15] Assiduous energetic works of greater depth and impact – such as, for example, the daily practice of a high level PENTA – can also lead to the stirring up of stagnated, self-contaminating energies.

evaluation of the practitioner, by a qualified measuring agent.

Periodically, comparisons between the participants' and the qualified measuring agents' measurements will have to be done, striving to identify patterns of agreement and discrepancy.

This research is planned to be conducted by this author. However, carrying out such a study and gathering data can occur only when a sufficient number of VELO practitioners reach a certain level of bioenergetic self-knowledge, parapsychic maturity, and energetic self-control. This aims to ensure that the body of research subjects/participants will have proficiency enough to satisfactorily fill out pre-designed forms dedicated to the investigation of their practical results.

It is expected that the training performed through the course *Goal: Intrusionlessness* will lead, in time, to a significant increase in the participants' self-mastery over the attributes discussed here as well as their acuity for bioenergetic self-perception and self-evaluation. This condition should produce a chain reaction, resulting in an increase in the number of members of the conscientiological community who have genuine bioenergetic self-control and knowledge, being capable, therefore, of producing true VSs – a requisite for the research plan described here.

After the accumulation of enough data, an article with the description of the results of this research study as well as pertinent discussions related to the results will be published.

Acknowledgments

I thank Wagner Alegretti for the time and patience dedicated to the revision of this work and for the suggestions and productive discussions about the expression of each VS attribute, which contributed appreciably to conferring a greater conceptual and redactional clarity to it. I also thank the physicist Massimiliano Sassoli de Bianchi for his critique and comments, which were fundamental in reaching a more precise and elegant interdisciplinary and mathematical expression of the presented theories.

Bibliography

Alegretti, W. (1992). Verbal communication. *Bioenergy: Theory and Practice* course. Instituto Internacional de Projeciologia; São Paulo, Brazil.

Alegretti, W. & Trivellato, N. (2005). *Bases for the Energogram and Despertogram*, PowerPoint presentation. Jornada de Despertologia. Centro de Altos Estudos da Conscienciologia. Foz do Iguaçu, Brazil.

Vieira, W. (1994). *700 Experimentos da Conscienciologia*. Rio de Janeiro, RJ. Instituto Internacional de Projeciologia e Conscienciologia, p. 348.

Vieira, W. (1999). *Projeciologia: Panorama das Experiências da Consciência Fora do Corpo Humano*. Instituto Internacional de Projeciologia e Conscienciologia. Rio de Janeiro, RJ, p. 384.

Note: This article was previously published in: Journal of Consciousness 11, No. 42, 2008, pp. 165-203. An Italian translation is also available in: *Lo Stato Vibrazionale*, AutoRicerca, Numero 1, Anno 2011, pp. 59-100.

AutoRicerca

From circular breathing to VELO: a proposal for an integrative technique

Massimiliano Sassoli de Bianchi

Issue 20
Year 2020
Pages 89-121

Abstract

After having emphasized some fundamental analogies between *Yoga* and the approach known as *Conscientiology*, we propose to combine the practice of a specific *pranayama*, called *circular breathing* (CB), with the technique of the *voluntary energetic longitudinal oscillation* (VELO), to promote a procedure as gradual and effective as possible in achieving the *vibrational state* (VS). We also present an ultra-simplified mechanical model to highlight one of the possible mechanisms underlying the functioning of these particular energetic (inner) technologies.

AutoRicerca, Issue 20, Year 2020, Pages 89-121

1 Introduction

Since the dawn of time men have asked questions about the nature and purpose of their existence. To search for non-speculative answers to these fundamental interrogatives, some individuals have developed, over the centuries, practical self-research methodologies aimed at achieving a possible clear vision not only of our material reality, but also of the spiritual (extraphysical) one. Probably, the oldest and most influential among the self-research methodologies, which has subsequently influenced every other approach, is that of *Yoga*. "Yoga" is one of the most well-known terms of the Indian cultural heritage, and nowadays it has also turned into a common word in western countries. In Sanskrit, the term derives from the root *yuj*, which means "to subjugate" and/or "to unite", and hints to the possibility for the human consciousnesses to master their numerous vehicles of manifestation, through a gradual and systematic integration of their different aspects, to awake their potential and promote an evolutionary acceleration.

The true origins of Yoga are difficult to determine, as in the ancient times its teachings have been handed down orally from masters to disciples, and nobody really knows for how long this oral tradition has lasted before the first known treatises were written. According to some authors, Yoga originated thousands of years before our common era, whereas according to some more speculative researchers its true origins are to be traced back to the presence of more advanced civilizations on our planet, in prehistoric times. But independently from its dating, there are no doubts that Yoga contains a complex corpus of high-level knowledge, not only about the anatomy and physiology of human beings, but of their psychology, para-anatomy and paraphysiology as well.

We can observe a number of significant parallels between the vision of man and cosmos contained in the ancient science of Yoga (as also expressed in the traditional Hindu scriptures, such as the *Upanishads* and *Vedanta*) and the one proposed in the ambit of the more recent approach to multidimensionality called

Conscientiology.[1] Let us give below three important examples.

Example 1. In Conscientiology, it is considered that the human beings are equipped with an entire holosoma, consisting of (at least) three intelligent vehicles (or bodies): physical (*soma*), emotional (*psychosoma*) and mental (*mentalsoma*). Similarly, in Yoga, it is considered that the individual consciousness (*atman* or *purusha*) is equipped with three bodies (*sarira*): gross (*sthula sarira*), subtle (*suksma sarira*) and causal (*karana sarira*).

Example 2. In Conscientiology, an *interface structure* of great importance is identified within the holosoma, called the *energosoma* (or *holochakra*): a complex matrix connecting the soma and the psychosoma, through which the latter can control and energize the former. Similarly, in Yoga the existence of an energosomatic interface is also recognized and named *prana maya kosha*, that is "illusory sheath (illusory in the sense of impermanent) made of energy."

Example 3. In Conscientiology, it is considered that every aspect of reality is the expression of three inseparable aspects (as are inseparable the three sides of a medal[2]): (1) energy, (2) emotion and (3) thought. For this reason, to describe the practical unit of manifestation of the consciousness, the neologism *thosene* has been coined, in which the root *"tho"* stands for "thought", *"sen"* for "sentiment" (also in the sense of emotion), and *"e"* for energy (also in the sense of matter). Similarly, in Yoga it is also recognized that the constitutive qualities of objects (*gunas*), namely the modalities through which energy takes shape and manifests, are three: *tamas* (matter, inertia), *rajas* (movement, activity) and *sattva* (cognition, intelligence), which correspond, respectively, to the three fundamental modes of perception: *sense, emotion* and *thought*.

The analogies between Yoga and Conscientiology may also be highlighted in their respective methods of self-inquiry and self-experimentation. Again, by way of demonstration, let us cite three

---

[1] According to its proponent *Waldo Vieira* (2002), Conscientiology is meant to be the science studying consciousness in an integral, holosomatic, multidimensional, multimillenary, multiexistential manner and, above all, according to its reactions with regard to immanent energy, consciential energy and its own multiple states.

[2] A medal has three sides: two are flat and one is curved.

important examples.

Example 1. In Yoga, the search for a perfect absence of movement is called *kaya sthairyam*, from "kaya", which means "body", and "sthairyam", which means "stability". In this technique, the practitioner, usually in a meditative posture, focuses solely on the possibility of keeping a full conscious stillness of his/her body, especially in order to experience a clear perception of the *prana* flowing into his/her physical and energetic vehicles. This practice is part of so-called *pratyahara* (fourth limb of Yoga, see next section), referring to that set of techniques that aim at inhibiting the ordinary sensory perceptions, as a prerequisite for having access to non-ordinary states of consciousness in the more advanced practices of concentration (*dharana*) and meditation (*dhyana*).

Similarly, in Conscientiology, it is recognized that the ability to master the stillness of the physical body is a prerequisite for the development of advanced parapsychic talents. Indeed, thanks to the conservation of immobility for a sufficiently long time (as it happens in the specific "consciential laboratory practice" named *waking physical immobility*, where perfect immobility of the physical body is ideally maintained for a time-period of 3 uninterrupted hours), the attention and awareness of the practitioner can easily disidentify from the ordinary physical dimension and expand beyond it, allowing him/her to access non-ordinary perceptions (paraperceptions) and enhance the development, among other things, of parapsychism and projectability.

Example 2. In Yoga there are several breathing techniques (pranayama) that include moments where the practitioner holds his/her breath between the phases of inhalation and exhalation. For example, in so-called *samavritti pranayama*, the practitioner seeks the arrest of normal mental functions and the acquisition of a state of mental stillness, by producing two phases of apnea, between the inhaling and the exhaling (retention with lungs filled), and between the exhaling and the inhaling (suspension with lungs empty), of equal duration as the phases of inhalation and exhalation (hence the name of *square breathing* often attributed to this pranayama).

The effectiveness of this procedure is due to several factors. Among these, there might also be that of reducing the amount of oxygen that normally circulates in the blood and increasing the amount of carbon dioxide, taking place gradually, without danger for the physiology of the human body and the neurophysiology of the brain. Indeed, this might reduce the efficiency of the physical brain, whose functions are therefore partially inhibited, although only temporarily, allowing the practitioner to more easily access paracerebral functions, associated to his/her subtler vehicles of manifestation.[3]

For the same reason, in Conscientiology a similar breathing technique is also used, although with only one phase of apnea between the inhalation and the exhalation, and a longer exhale with respect to the inhale. This technique (called *triangular breathing* in Yoga) is called *carbon dioxide technique* in Conscientiology, and is considered one of the most effective in producing lucid extracorporeal experiences, when it is practiced in the supine "corpse" position (*savasana*), with the body perfectly relaxed; see for instance (Vieira, 2002), p. 448.

Example 3. In Conscientiology, great importance is attached to the different bioenergetic training techniques. Similarly, in Yoga it is considered that the techniques of activation and control over the different energetic fluxes, usually referred to as *kriya* (a term that means "act" or "action", and refers to the practical aspect of something, as opposed to the theoretical one), *pranayama* (control over *prana*) and *pranavidya* (knowledge of *prana*), are absolutely indispensable for the practitioner's inner progress.

Purpose and organization of the article

The main purpose of this paper is to consider more closely this latter parallel between Yoga and Conscientiology, by proposing to combine the practice of a specific *pranayama* of Yoga with that of a specific "conscientiological technique": the *voluntary energetic longitudinal oscillation* (VELO) technique; this in order to promote a methodology of practice that can possibly help the practitioner to reach, in a step by step manner, and independently of his/her initial level

---

[3] Note however that according to so-called *Bohr effect* (named after the Danish physiologist *Christian Bohr*, who described it in 1904), a reduction of the intake in air can also promote (particularly for people having the tendency to hyperventilate) a better vasodilation and a better oxygen transport in vital organs. In other words, breathing less does not necessarily mean oxygenating less.

of discernment of bioenergy, a full mastery of so-called *vibrational state* (VS): a condition of maximum energosomatic (and possibly holosomatic) dynamization, which is at the basis of many of the paraphenomenologies of the consciousness.

The paper is organized as follows. In the next section we briefly introduce the reader to the concepts of *pranayama* and *pranavidya* of Yoga. We will then describe in detail the classic *ujjayi pranayama* (UP) technique. Subsequently, we will explain the (much less known, also in the ambit of Yoga) technique of *circular breathing* (CB). After that, we will summarize the essentials of the VELO technique and then present our proposal for an integrative technique, named VESELO, which will combine the practice of the CB (which, in turn, is based on the UP) with the VELO. Finally, we will discuss the pros and cons of the VESELO, as compared to the VELO, and offer some concluding remarks.

In the appendix of this work, we also describe an ultra-simplified mechanical model, illustrating a possible mechanism of action of the consciousness on the energosomatic fluid, which could be one of the ingredients explaining the functioning of both the CB and VELO techniques.

2 Pranayama and pranavidya

The most distinguished master of Yoga of the past is undoubtedly *Patanjali*, whose identity and dating are uncertain. Patanjali, in his famous *Yoga Sutras* (Patanjali, 2003), bequeathed what is today considered the most authoritative text on the subject of Yoga. In this writing, the author didactically breaks up the yogic practice in 8 (distinct, but intertwined) limbs, or stages, which is the reason for the term of *Astanga Yoga* (*asta* = eight, *anga* = limb), sometimes used to describe Patanjali's system:[4]

1. *Yama* (the don'ts, the *abstinences* from all that impedes evolution; to be also understood as a social behavioral code);

---

[4] The classical Patanjali's eightfold path of Yoga is also called *Raja Yoga*, that is, *Royal Yoga*.

2. *Niyama* (the do's, the *observances* of all that promotes evolution; to be also understood as a personal behavioral code);

3. *Asana* (the mastering of a stable and comfortable *posture*, both physical and mental, both external and internal, to be also understood as a state of being);

4. *Pranayama* (*control* and regulation of the respiratory and energetic fluxes, of the life force, of the immanent and consciential energies);

5. *Pratyahara* (*"inward" focus*, by the withdrawal of the ordinary senses);

6. *Dharana* (*concentration* of the individual self on a single object);

7. *Dhyana* (*meditation*, mindfulness, uninterrupted attention to the object of concentration);

8. *Samadhi* (*cosmoconsciousness*, fusion without confusion of the individual self with the universal principle, transcendental consciousness).

As stated in the introduction, in the present work we are more particularly concerned about the fourth limb of classical Patanjali's Yoga: the practice of *pranayama*. The word "pranayama" means "control over breath" (*prana* = breath, inhalation, immanent energy; *ayama* = control, extension, expansion) and indicates the entire set of procedures aiming at the control and direction of the physiological breathing (and, more generally, of the bioenergies) linked to the practice of Yoga.

We can observe that the breath – that is, the gas exchanges between the inside and the outside of our physical body (*exteriorizing* and *interiorizing*), as well as those taking place only internally, between its various organs and systems (*internal circulation*) – constitutes one of the subtler forms of *energetic mobilization* that a human consciousness is able to execute and monitor through the somatic instrument and the associated ordinary perceptions. In that sense, the physiological breath is a kind of last frontier, still perceptible and usable through our ordinary senses, beyond which unfold the dimensions of *extraphysical* nature, domain of manifestation of our subtler consciential vehicles.

From this simple observation, one can understand the importance attributed to the breath in the practice of Yoga: the regular application of pranayama's techniques (in combination with the *asanas*) allows not only to oxygenate the entire biological organism (particularly the brain and the nervous system), but also to act at the interface between the physical and the extraphysical, unblocking and releasing those more subtle consciential energies that are at the basis of our physical and nonphysical manifestation. In Yoga, the breath plays in fact a double role: a physical one, associated to the puff and blow movement of the lungs, and a "psychic" one, associated to the energetic movements of the subtle bodies, especially the energosoma.

For practicing the energetic "respiration," there is, in the ambit of Yoga, a vast corpus of paratechnologies (inner technologies), generically referred to as *pranavidya* (which means "knowledge of prana", in Sanskrit), whose purpose is helping the practitioner promoting the full development and mastery of his/her energetic sphere. In the Yoga Sutras, Patanjali makes a clear reference to the possibility of moving our energies also independently of our control over the physical respiratory fluxes. More precisely, says Patanjali (in sutras 50-52 of the book of practice), in addition to the usual control of the three phases of inhalation, exhalation and suspension/retention (*puraka*, *rechaka* and *kumbhaka*), there is also a *fourth kind of control*, which is no longer the respiratory act as such, or its suspension/retention, but rather the direct control over the pranic currents that run through the *prana maya kosha* (energosoma).

This fourth control consists in the possibility of shifting the perceptive level of the practitioner from the *anna maya kosha* (another Sanskrit term for the physical body, which means "illusory body made of food") to the *prana maya kosha*, by directly and consciously moving the energies of the latter along certain passageways and/or centers (such as the *nadis* and *chakras*). When such an energosomatic breathing is executed and controlled, says Patanjali, the physical body remains as inert as possible, quiescent, relaxed, allowing the practitioner to experience expanded states of consciousness and an increased level of lucidity.

Therefore, in general, pranayama's techniques are meant to move the energies not only within the physical body, via the regulation of the gas fluxes, but also within the energosoma, and beyond, via the control of the subtler "winds" (fluxes). This is done by promoting a practice of conscious breathing, where the different respiratory moments are accompanied by a specific mental focus, as well as by promoting a purely energetic "breath," triggered only by the intentionality and will power of the practitioner.

It should be noted that in Yoga there is a very thorough and detailed description of the different types of pranic fluxes, which, depending on how they move and go through the complex circuitry of the energosoma, along *nadis*, *chakras* and other energy centers, are transformed, structured (i.e., conscientized), assuming different properties. The tradition distinguishes *ten* different types of pranic fluxes, according to their movements and directions, five of which are considered to be the most important:

1. *prana* (inhalation);
2. *apana* (exhalation);
3. *samana* (assimilation);
4. *udana* (expression);
5. *vyana* (distribution).

Of these five, the first two, *prana* and *apana*, are considered to be truly fundamental. Typically, *prana vayu* (*vayu* = wind) is the *ascending flux*, associated to inhalation (the term *prana* also means to inhale) and interiorization of energies necessary to the sustainment of life, while *apana vayu* is the *descending flux*, associated to exhalation and exteriorization of energies, to be also (but not only) understood as the elimination of energies that are no longer usable.

The fundamental alternation between these two *ascending* and *descending* movements is called *prana-apana-gati* (*gati* = path, speed). In *pranayama*, through the alternation of the *prana-apana-gati*, one tries not only to move the energosomatic substances to increase their fluidity, but also to bring together and unite these two fundamental "winds", especially at the switch points between the ascending and descending movements.

3 Ujjayi Pranayama

Let us now start considering the more practical aspects of this work. We will begin with a synthetic, but nevertheless comprehensive, description of the fundamental technique of the *ujjayi pranayama* (UP): a breath control procedure that promotes a full chest expansion, in an upward movement of "conquest" (the term "ujjayi" is the composition of "ud," a prefix meaning upward, dilated, and "jaya," meaning conquest, victory).

Posture. The technique can be practiced in different body positions. What's important is that the back is maintained upright and that the abdominal area is relaxed. Typically, we can mention the following four possible positions of practice:

1. standing (*tadasana*);
2. lying on one's back (*savasana*);
3. meditative posture (e.g., *swastikasana, sukhasana, siddhasana, padmasana, virasana*);
4. sitting on a chair (the back will then be self-sustained and preferably not supported by the back of the chair, and the soles of the feet will be kept in contact with the ground).

Technique. It entails the execution of the following *4 stages* (the first two stages are only preparatory for the third, which corresponds to the technique as such, and can be skipped by the experienced practitioner; the fourth stage, of observation, is only optional, although recommended):

Stage 1. For a time, bring your attention to your natural breathing, *only through the nostrils*, noting the expansion of the abdomen during inhalation phase and its contraction during the exhalation phase.

Stage 2. Slowly, extend the two respiratory phases and through the partial closure of the glottis (as when one clears the throat, or whispers) make perceptible the air flow passing through the base of the palate, producing a characteristic *hissing sound*, similar to the backwash of the sea (the breath is always and only from the nose). The sound must be subtle, but nevertheless loud enough to be heard by the practitioner (also, it has to be uniform and pleasant to hear). Do this for a certain number of respiratory cycles.

Stage 3. Subsequently, by applying an active control over the breath, *fill the lungs from the bottom upwards, during the inspiratory phase, and empty them from the top downwards, in the expiratory phase.* To this end, an instant before breathing in, exert a gentle but firm contraction of the lower abdomen, pulling it back toward the spine, then lifting the belly up, promoting in this way a full upward expansion of the chest. During the exhale, keep at first the chest lifted and the abdomen contracted, then lower first the shoulders and let the chest gradually close and come down, finally releasing the abdomen (in other words, the contraction and expansion of the abdomen is opposite to the movement of the physiological breath[5]). Do this for a sufficient time of practice, trying to make the inhalation and exhalation of equal length, intensity and as uniform as possible.

Stage 4. After having abandoned the technique, just observe the condition that has been reached, taking note of the changes that occurred in the physical, energetic, emotional and mental spheres.

Observations. The particular hissing sound produced during the UP cannot, for obvious reasons, be fully grasped by just reading a written explanation. It is therefore important to hear it as is being executed by a practitioner who is familiar with the technique.

Initially, try to avoid creating too many tensions when applying the technique, which may appear a bit difficult, especially with regard to the contraction and expansion of the lower part of the abdomen.

During the exhalation phase, the relaxation of every muscle, especially of the shoulders, neck and face, promote a slight movement of the head towards the trunk. Conversely, during the inhalation phase, the head will gently rise, as though it would extend the movement of expansion of the chest.

The inhalation and exhalation should be *continuous* and *uniform* in their unfolding, without any hurry, or shortness of breath; also, they should be at the same time *light* and *deep*, with the *ujjayi* sound barely audible from the outside.

When practicing the UP, we can observe that during the phase of inhalation it is possible to perceive an energetic current propagating

---

[5] The contraction of the abdomen in the initial phase of the inhalation, and its expansion in the final phase of the exhalation, is also known as *prenatal breathing*. In fact, when we were in our mother's womb, though not exchanging oxygen with the external environment (being completely immersed in the amniotic fluid), we absorbed nutrients and oxygen through the umbilical cord, in a continuous exchange with the maternal organism. This exchange can be likened to a sucking movement (absorbing energy) promoted by the contraction-compression of the abdomen, and to an opposite out-thrust movement (elimination) promoted by the expansion-decompression of the abdomen. Therefore, when practicing the prenatal breathing, we evoke again this ancient state of symbiosis with the maternal organism, which fully supported and protected us, in an experience of profound unity.

not only from the outside in (interiorization of immanent energies, especially those contained in the air), but also from the bottom to the top (from the pelvic floor to the top of the skull), along a path parallel to the spine; this even though the physical breath of the air entering the lungs corresponds to a filling movement which is in fact opposite, as descending.

Likewise, during the exhalation phase, it is possible to perceive a current of energy propagating not only from the inside out (exteriorization), but also from the top to the bottom, i.e., from the top of the head to the pelvic floor, in accordance to the downward direction of the *apanic flux*.

These movements arise as a result of the successive movements of contraction of the abdomen and expansion of the rib cage, during the inspiratory phase of the ujjayi pranayama, producing a volumetric upward movement of expansion, in accordance with the direction (usually) perceived of the ascending *pranic flux* (and vice versa in the expiratory phase).

In this regard, we can observe that according to the different traditions, in the region of the abdomen a "reservoir of vital/sexual energy" would be located, an "ocean of chi", called *dantian* (cinnabar field) in the Chinese tradition, *hara* (belly, abdomen) in the Japanese one, and *kanda* (bulb) in the Indian one. In this region, which is located about three fingers below and two fingers behind the navel, the basic life energy of the individual would accumulate and be distributed.[6]

Thanks also to the contraction of the abdomen during the inhalation, the energy contained in this reservoir would be pushed upwards, going then to meet the incoming pranic flux, inhaled with the air (and vice versa in the expiratory phase). As a result, the meeting and mixing of these energies would be able to produce an enrichment and refinement of the overall energosomatic energies, thereby increasing the amount of energy mobilized by the respiratory process.

The UP can be practiced at any time of the day, without any specific contraindication. However, one must take care not to execute the technique mechanically, that is to say with unawareness, since

---

[6] According to the Chinese tradition, there would be two other reservoirs of energy in the para-anatomy of the individual: one located in the middle of the chest and the other in the middle of the forehead, corresponding to more subtle energetic qualities.

the risk in this case is to create tensions in the respiratory dynamics. We must always remember that altering the natural breathing flow, without the assumption of a correct mind posture, and the necessary relaxation, can produce in the long run more negative than positive effects.

4 Circular Breathing

Circular breathing (CB) is a technique of breath control in which the phases of retention and suspension (apneas) are completely abolished. It is a very powerful respiratory procedure, described in some ancient texts (not just in Yoga, but also in other traditions, such as the *Sufis*). For example, we find traces of it in the following passage from the *Bhagavad Gita*:

"[...] some others, engaged in the practice of pranayama, by regulating the incoming and outgoing breaths, offer the inhalation (Prana) into the exhalation (Apana), and the exhalation (Apana) into the inhalation (Prana)."

Also, in the *Vigyan Bhairava Tantra*, we find mention of this technique in the following passage:

"Thanks to the collision of the two vital breaths, inside or outside, the yogi enjoys, in the end, the birth of the consciousness of sameness."

In the CB one tries to produce a collision, of an explosive type, between the inhalation and the exhalation fluxes, and vice versa, so that even before a respiratory phase is being completed, the other one already takes over. The contraposition of these two linear and complementary processes, merging one into the other, then produces a *circular-like* dynamic, as exemplified by the *Tai Chi* symbol of the Chinese tradition (hence the name of this procedure).

Posture. Take one of the four positions described in the previous technique.

Technique. It entails the execution of the following *6 stages* (the first four stages are just a preparation for the fifth, which corresponds to the technique as such, and can be skipped by the experienced practitioner; the sixth stage, of observation, is only optional, although recommended):

Stage 1 to **Stage 3**. Same as in the previous technique.

Stage 4. Always breathing only through your nose, extend the inhalation and exhalation phases, which should remain of equal duration, bringing them up to about *7 seconds*.[7] The breath (and the corresponding sound) should be rendered as uniform as possible, without producing accelerations or bursts, simply observing its U-turn points, where the inhalation, once exhausted, gives way to the exhalation, and vice versa. Proceed in this way for several breathing cycles.

Stage 5. Then, apply the circular breathing technique per se: rather than allowing the inspiratory phase to come to an end, at about *3/4* (three quarters) of it, produce a sudden reversing of its direction, without any transition, passing in an "explosive" manner from the inhalation to the exhalation; same thing for the expiratory phase, which must also be stopped at *3/4* of his path, to be instantly replaced by the inspiratory phase, and so on. In this way, the duration of the two phases will be slightly shortened, falling to just over *5 seconds*. Continue to alternate in this way inhalation and exhalation (only through the nose and using the UP technique), for a sufficient time of execution of the technique.

Stage 6. After having abandoned the technique, just observe the condition that has been reached, taking note of the changes that occurred in the physical, energetic, emotional and mental spheres.

Observations. In the beginning, it may be preferable to practice the technique lying supine, and only afterward, once acquired a certain expertise, practice it in the sitting or standing position. The standing position requires some caution, if one doesn't have the necessary experience, in order to avoid dizziness that may cause dangerous falls.

During the application of the technique there is the tendency to gradually shorten the duration of the two respiratory phases. To avoid this, try to keep the rhythm as steady and regular as possible, avoiding becoming anxious.

---

[7] This figure is of course only indicative, since the ability to extend the respiratory phase is a function of the experience and level of practice of the practitioner, as well as of the specific anatomy of his/her lungs.

The crucial aspect of the technique is in the explosive switches between the two respiratory phases, producing at each inversion a small *energetic shock*, able to increase both the frequency and intensity of the subtle energetic flux that accompanies the breath.

This technique produces a significant energetic dynamization of the *soma* and *energosoma* of the practitioner, thus promoting the release of tensions, as well as energetic and emotional blockages; if this happens, just let the emotional energy flow, without blocking it or promoting unnecessary identifications (keep an attitude of neutral and detached observation).

The application of the technique, especially if extended beyond *fifteen minutes* and executed with intensity, can produce effects of *muscle tetany*, due to the low levels of carbon dioxide in the blood. These effects are not dangerous and will quickly disappear at the end of the breathing.

At the end of the practice, it is possible to experience a *natural breath suspension*, due to the very low levels of carbon dioxide in the blood stream and the deep energization that occurred during the application of the technique. It is desirable to take advantage of this moment to experience the profound sense of stillness and liberation promoted by the cessation of the respiratory mechanism, which is typical of extracorporeal states.

One should not confuse the CB technique described above, with the breath-works described in other practices (such as *rebirthing*), where the cyclic breathing with no pauses between the inhale and the exhale is often performed through the mouth, in a very uneven and unbalanced way, without the use of the UP, and the inner posture of the practitioner is totally passive, with all that that it implies in terms of lowering of the natural energetic defenses and possible exposure to all forms of negative subtle influences.

5 Voluntary energetic longitudinal oscillation

In this section we will illustrate the key elements of the *voluntary energetic longitudinal oscillation* (VELO): a technique of control and mobilization of the bioenergies (i.e., a technique of energetic "breathing" of the energosomatic parabody, and not of physiological breathing of the soma). The main purpose of the VELO is to promote the so-called *vibrational state* (VS), a particularly intense condition of energetic dynamization and activation, that can unlock many evolutionary possibilities (usually perceived as a vibration diffusing throughout the whole body and parabody, able to sustain itself for a certain time after the application of the technique).

We can observe that in the vast corpus of Yoga's techniques, related to the knowledge of prana (pranavidya), there are a number of methodologies of energetic mobilization, some of which are quite similar to the VELO. For instance, the *prana-apana-gati* itself, that we already mentioned, if not directed to the somatic breath, but to the pranic energies, corresponds in fact to a periodic (oscillatory) longitudinal movement of the energies. Another example is the so-called *sthula-bedhana kriya* (literally: action of "piercing" the body, in the sense of passing through it with the pranic energies), where the energetic longitudinal movement is accompanied by a further upward exchange of energy, through the palmochakra of the two arms up (in the direction of the sky), and a downward exchange of energy, through the plantochakra of the feet resting on the ground (in the direction of the earth).

We can therefore consider that the VELO technique is, to some extent, a variant, or a modern reworking, of an ancient yogic technique of energetic sweeping. As far as this author knows, in its present form the VELO technique is due to *Waldo Vieira* (2002),[8] although we can find traces of it in the first book written by another modern pioneer of lucid projection: *Robert Monroe.*[9]

---

[8] The first edition of the projectiological treatise of Vieira dates back to 1982.

[9] In his first book (Monroe, 1977), firstly published in 1971, Robert Monroe proposes, on page 211, a completely different technique from the VELO, to reach the VS. It's a purely mental procedure, which is akin to yogic techniques of concentration on the inner mental space, which can be perceived in front of the closed eyes, just behind the forehead, called *chidakasha*. However, Monroe also proposes a technique of *control of the vibrations*, to be used once the VS has been installed, which in fact is quite similar to the VELO, that subsequently has been better defined and described by Waldo Vieira. More specifically, on page 214 of (Monroe, 1977), we can read the following description: *"First, mentally 'direct' the vibrations into a ring, or force them all into your head. Then mentally push them down along your body to your toes, then back up to your head. Start them sweeping in a wave over your*

This author learned for the first time about the VELO (or "VS exercise") in year 2000, in a channeling session with the American medium *Jon C. Fox*.[10] Later on, he learned about the technique directly from the courses offered by the *International Academy of Consciousness* (IAC), and more specifically from the writings of Waldo Vieira (Vieira, 1994), p. 348, (Vieira, 2002), p. 587, (Vieira, 2003). More recently, the VELO technique has been explained, at some length, in an article by *Wagner Alegretti* (2008), and further thoroughly analyzed by *Nanci Trivellato* (2008). For completeness, and for the commodity of the reader, we present below, once again, the fundamental elements of the VELO technique.

Posture. Take one of the four positions described in the previous techniques.

Technique. It entails the execution of the following *5 stages* (the

---

body rhythmically, from head to toes and then back again. After you have gained the wave momentum, let it proceed of its own accord until it fades away. It should take about ten seconds – five down, five back – for the wave to make the complete circuit, from head to toes and back. Practice this until the vibration wave begins instantly upon mental command, and move steadily until fade-out. By this time, you will have noticed the 'roughness' of the vibrations at times, as if your body is being severely shaken right down to the molecular or atomic level. This may be somewhat uncomfortable, and you will feel a desire to 'smooth' them out. This is accomplished by 'pulsing' them mentally to increase their frequency. [...] Your first indication of success is when the vibrations no longer seem rough and shaking. You are well on your way to control when they produce a steady, solid effect. It is essential that you learn and apply this speed-up process. The faster vibration effect is the form that permits disassociation from the physical."

[10] During the session, the presumed extraphysical entity known by the name of *Hilarion*, synthesized as follows the method to achieve the energetic condition of the VS: *"Vibrational state exercise is done by feeling, really creating deep in your consciousness full awareness, not just visualizing or imaging but really feeling the presence of energy. Begin by the awareness of the energy in your head. Move it slowly through the body to the feet. Reverse the direction back up to the head. If you are seated, move the energy in a diagonal line, not moving along the contours of the body, but simple straight diagonal line. If you are standing or lying down, move it in a straight line. Simple exercise, ay? Then, as you move the energy, and encounter any blockages, simply increase the energy. Move it stronger through those areas of blockage. Then move the energy faster, up and down. Faster, faster still, faster than you can imagine, even faster still. Do not hold the body. Do not hold breath. Do not tense anywhere. No physical manifestation of the energy is necessary. Very simple exercise. Do it repeatedly during the day. A field of energy is gathered around you as a result of this that has tremendous value on many levels, keeping away the influences of the nonphysical beings of a lower vibrational nature, naturally welcoming and opening to the vibrational energies of those of a high, helpful nature. And many other things this does."*

first stage is just preparatory and can be skipped by the experienced practitioner; the fifth stage, of observation, is only optional, although recommended). We will describe the technique assuming that the practitioner sits in a meditative posture, for example *siddhasana*, and will explain afterwards what should be considered when practicing in other positions.

Stage 1. Remain still in the posture, and just bring your attention to the whole of your energies (energetic sphere), trying to perceive at best their presence and quality.

Stage 2. Using your will, try to concentrate the greatest possible quantity of bioenergy in the region of the head [it is also possible to start from the region of the feet, or from another region of the body; see the discussion in Trivellato (2008)].

Stage 3. Once a sufficient concentration has been reached, move the energies[11] localized in the head downwards, longitudinally, along the main axis of the body, to the base of the pelvic floor, then reverse the direction of propagation and direct the energies upwards, to again reach the top of the head, then down again, and so on, slowly sweeping (scanning) up and down the entire corporeal volume. Continue in this way for a certain time, moving the energies slowly and deeply through each longitudinal section of the body.

Stage 4. Increase the power of the energetic movement, trying to gradually increase, within the limits of your paramotor skills, both the quantity of energy put into motion by the mental action, and the frequency of the alternating current running throughout your soma (i.e., the speed of the scanning longitudinal movement). Do this trying to keep the body perfectly relaxed, without going out of rhythm, or losing the depth and amplitude of the sweeping movement, making sure that the inversion of the direction of the energies in the two extreme points of the movement takes place instantaneously, without slowdowns or pauses. Do this for a sufficient time of execution of the technique.

Stage 5. After having abandoned the technique, just observe the condition that has been reached, taking note of the changes that

---

[11] The term "energies" should also be understood here as "subtle energetic substances."

occurred in the physical, energetic, emotional and mental spheres.

Observations. If you practice in the standing (*tadasana*) or supine (*savasana*) positions, with the arms alongside the body, the sweeping (scanning) movement of the energies must affect the entire length of the soma, from the top of the head to the soles of the feet, passing through the trunk and legs, and vice versa.

If, on the other hand, you practice sitting on a chair, in the third stage of the technique, when the energy is still moving slowly, just follow the contour of the body, to ensure the passage of the energy through every longitudinal section of the body. Then, in the fourth stage, when and if the energetic scanning movement accelerates and becomes very rapid, being no longer possible to exactly follow the contours of the body, simply move the energy in a straight line, along a diagonal.

A crucial aspect in the execution of the VELO is the ability to perceive and direct the energosomatic substances for real, and not only to imagine doing it. By moving the mental focus along the axis of the body, it is certainly possible, in part, to put these "subtle" substances into motion, however, to develop an energetic movement of sufficient power, able to install the VS, the practitioner has to learn, with time, to develop a true paramotility, namely the ability to effectively direct the bioenergies, in the same way s/he is able to direct, for example, her/his physiological breath, or the movement of a limb of the physical body.

There would be much still to be added to the analysis of this particular paratechnology, which is simple to execute only in appearance. For this, we refer reader to the aforementioned writings of *Trivellato*, *Alegretti* and *Vieira*.

6 Voluntary energetic somatic and energosomatic longitudinal oscillation (VESELO)

In this section, we present our proposal for an integrative technique, which considers the combination, and in part the blending, of the CB somatic technique with the energosomatic technique of the VELO. We shall denote this mixed technique the *voluntary energetic somatic and energosomatic longitudinal oscillation* (VESELO).

Posture. Take one of the four positions described in the previous techniques (supine, standing, sitting on a chair, or in a meditative posture).

Technique. It entails the execution of the following *4 stages* (the first stage is essentially preparatory and can be skipped by the experienced practitioner; the fourth stage, of observation, is only optional, although recommended):

Stage 1. Practice the CB, until reaching a significant mobilization of the somatic and energosomatic energies, by maintaining, for long enough, a uniform and relatively intense respiratory rate. In this stage, focus only on the proper execution of the UP and on the explosive inversions between the phases of inhalation and exhalation.

Stage 2. Pair the somatic breathing with a conscious energosomatic breathing. More precisely, in the inspiratory phase, focus the attention on the ascending *pranic flux* (in accordance with the upward movement promoted by the UP), from the base of the column (if you are sitting in a meditative posture), or from the base of the feet (if you are sitting on a chair, standing or lying down), to the top of the head. Conversely, during the expiratory phase, focus on the descending *apanic flux*, from the top of the head to the base of the column (or feet). Continue in this manner for a sufficient time, trying to couple in a synergistic way the physical breath control with the energetic movement of the *prana-apana-gati* associated with it, trying to make it as intense as possible.

Stage 3. Once a sufficient stability in the double oscillation of the respiratory and energosomatic fluxes has been reached, without interrupting in any way the movement of the energies, abandon the CB technique (i.e., the active control on the breath) and concentrate solely on the energosomatic mobilization, following the procedure described in stage 4 of the description of the VELO, trying to gradually and uniformly increase the power of the energetic flux. Do this for a sufficient time of execution of the technique.

Stage 4. After having abandoned the technique, just observe the condition that has been reached, taking note of the changes that occurred in the physical, energetic, emotional and mental spheres.

Observations. In the second stage of application of the technique, it is important to maintain the energetic sweeping perfectly synchronized with the somatic respiration.

The transition between the second and third stage must take place without interruptions or bumps in the sweeping rhythm of the energies. It is important that in this delicate passage the practitioner doesn't lose his/her mental focus on the energies in motion. Indeed, until that moment, the energetic flux was primarily sustained by the action of the breath, and only secondarily by the mental action. After that moment, i.e., in the passage from the second to the third stage, as in a relay race, there is an "exchange of the baton": the breath ends its primary directive action and, so to say, abandons the race, while the mental action totally assumes the control over the energetic mobilization, producing a gradual acceleration and further intensification of the same.

7 Advantages and disadvantages of the VESELO

An obvious disadvantage of the VESELO, compared to the VELO, is that it's a hybrid, more complex technique, requiring the control not only of the extraphysical "winds," but also of the physical (respiratory) ones. Therefore, in that respect, its approach is less immediate than the VELO.

On the other hand, the effort and time required to achieve a sufficient level of control in the practice of CB, is amply repaid by the fact that, thanks to the synergistic action between the CB and the VELO, it becomes possible, typically, to set in motion a more significant quantity of energy.

The VESELO technique has the disadvantage that it could lead the inexperienced practitioner to believe that there is a link between the physiological breath and the energosomatic "breath." In that respect, the practice of the VESELO could delay the development of the capacity to act directly on the subtler parts of the energetic sphere, and therefore master the VELO as such.

To avoid this possible misunderstanding, it is important that the instructor, during her/his theoretical introduction, gives due emphasis to the fact that there is no constraint, but only a possible synergy, between the physical "breath" and the energetic "breath." It is also important for the practitioner to fully understand what the logic of the VESELO technique is. It merely seeks to exploit the breathing techniques as a tool to indirectly set in motion a sufficient quantity of "subtle" energy ("subtle" substances), starting from the mediating action of the breath, which acts at the frontier between the soma and the energosoma. In other words, the first two stages of the technique are only preparatory for the third, where the energy is guided solely by the mind action of the practitioner.

An expert practitioner can of course do without this preparation, as s/he is already able to operate with sufficient paramotor effectiveness on her/his energosphere. Thus, while progressing in the practice, the tendency will be to increasingly reduce the first two stages and extend in proportion the duration of the third, thus arriving, in the end, at the VELO practice in its "classic" or "pure" form.

A disadvantage of the VESELO technique is to shift the focus of the practitioner on the physical dimension, through the control of the breath, whereas the primary objective is to learn to act directly on the extraphysical energies. Therefore, its practice may become counterproductive in the long run.

Any strategy that aims to obtain a specific result must be used with the necessary intelligence and discernment, otherwise, as is known, the solution of a problem, if acted in an inappropriate manner, can become part of the problem itself. The main difficulty in the proper execution of the VELO is the notorious lack of a clear perception, by the practitioner, of her/his energosomatic energies, and their discrimination with respect to the somatic ones. Another major difficulty is the observed viscosity (lack of fluidity) of the energosomatic energies in a number of practitioners (especially among those who have not passed through, in their previous intraphysical life, the second death process), a factor which makes them very difficult to mobilize.

This is perhaps one of the reasons why Yoga, whose corpus of methodologies offers a systematic and gradual approach to

consciousness' expansion (practicable regardless of the initial evolutionary level of the practitioner), gives so much importance to
bodywork (*asana*) and breathwork (*pranayama*), as a *sine qua non* condition for becoming then able to act in an effective way on the subtler spheres. Indeed, one of the goals of the asana and pranayama
practice is precisely to promote an intense unblocking and fluidiza-
tion of the physico-energetic sphere of the practitioner, enabling
her/him to explore and act more easily and effectively on her/his
more subtle anatomy and physiology.

From that perspective, the integration of the CB in the VELO
practice appears to be fully functional in bringing the practitioner
towards a greater awareness of his/her own "holosomatic machine" and develop the capacity to act directly on it, with greater
efficiency and effectiveness.

As is known, a crutch is useful in a process of rehabilitation of a
limb that remained motionless for too long only to the extent that,
at the appropriate time, it is abandoned, in order to allow for a full
recovery. Similarly, the "breath crutch" should be used only until
the removal of the main obstacles that impede the proper flowing
of the pranic energies; once this condition has been reached, it must
be abandoned, in the sense that the practitioner must learn to completely disengage his/her bioenergetic practice from the bellows'
mechanism of his/her lungs.

Of course, this doesn't mean that the pranayama practice as such
will then lose all its *raison d'être*. Indeed, these methodologies are
particularly effective in their action on our denser energies – physical and quasiphysical – and therefore can always be used for their
beneficial dynamizing and decongestant effects on our organism
(as it would be the case, for example, for a detoxifying and nutri-
tious diet, or for other practices that can promote a superior health
and hygiene).

An element of indubitable interest in the VESELO practice is the
phase of natural breath suspension that occurs at the end of the CB
(proceeding from the second to the third stage), when the active
control of the breath is abandoned. In this particular moment, the
practitioner doesn't feel the need to breathe for a period of time
which is usually fairly long, thereby reducing the interferences produced by the usual lungs' activity. This condition of respiratory immobility obviously favors a greater concentration on the

energosomatic action.

We also note that if one practices in the supine position, the breath suspension, combined with the massive mobilization of energies produced by the CB, further intensified and refined in the third stage of the VELO technique per se, undoubtedly constitutes in itself an effective technique for the lucid projection of the consciousness.

In the passage from the second to the third stage of the technique, particularly if the second stage has been practiced for long enough, it is quite easy to perceive the oscillatory movement of the energies that have been put into motion by the joint action of the breath and mental focus. This perception, which is rather concrete, is of considerable help for the less experienced practitioners, who thus have access to more objective paraperceptions, which can in turn facilitate the purely mental mobilization of the energies.

8 Conclusion

The author hopes that the present work will help foster a fruitful dialogue. Firstly, by stimulating modern researchers and self-researchers not to underestimate the value of the information contained in the ancient teachings of Yoga, which, in all likelihood, are the legacy of fairly advanced consciousnesses who walked this planet in the past millennia.

Secondly, by stimulating the serious Yoga practitioners not to commit the mistake to confuse antiquity with authoritativeness. In fact, the information contained in the ancient texts are not necessarily always correct and, as a rule, it should be translated into a language as clear and scientific as possible, avoiding unnecessary dogmatisms and pomposities, being aware that many of the yogic techniques need to be reviewed (without of course distorting them) by taking into account the different condition of modern man.

In other words, this author hopes that it will be possible to build a bridge across the different traditions of inner research that appeared over the millennia on this planet, to better understand their common foundations and possibly integrate (for the better) their

paratechnologies. However, this should be done with due serious-
ness and solely for the purpose of enhancing the effectiveness and
efficiency of the evolutionary tools at our disposal, and not to look
for easy paths of least resistance, inspired by self-corrupting mechanisms.

That said, we emphasize that in the description of the UP, CB
and VESELO techniques, we have voluntarily omitted to state the
specific times of practice (both for the techniques as such and for
each of their stages). Similarly, no specific guidance was provided
on the levels of respiratory intensity required. This mainly for
three reasons:

1. Every practitioner should be able to monitor and determine
    for themselves the right intensity and duration of application
    of the different breathing techniques;
2. It is not possible to delegate to a simple article the delicate task
    of guiding a beginner in the practice of these procedures. It is
    therefore recommended to initially practice them under the su-
    pervision of a more experienced practitioner;
3. A certain prudence is necessary when using respiratory proce-
    dures of a certain power, which, if applied recklessly and inap-
    propriately, can induce uncomfortable phenomena, and some-
    times even dangerous ones, such as respiratory alkalosis, hy-
    perventilation, hypoxia, nausea, vomiting, paresthesia, arrhyth-
    mias, panic attacks, etc. If we stress all this is not to promote
    undue fears in the practitioner, but simply to remember that it
    is essential to execute the techniques with due discernment,
    knowledge, and a clear perception of one's limits.

It is well known that many other Yoga's pranayama techniques
are able to promote deep dynamization of the energetic sphere
and possibly lead the practitioner to the installment of so-
matic/energosomatic vibrational states (Alegretti, 2008), (Sassoli
de Bianchi, 2018). We can cite for instance the famous *bastrika
pranayama* and *kapalabhati pranayama*, or the so-called *breath of fire*.
However, these procedures, for reasons it would be too long to
explain here, are not suitable to be harmonically integrated with
the VELO technique.

The CB pranayama (not to be confused with the *Rebirthing* practice
by *Leonard Orr*, or the *Holotropic Breathwork* by *Stanislav Grof*, despite

some obvious similarities[12]) undoubtedly has points in common with the VELO. Firstly, in the importance of creating, when applying the technique, a stable rhythm and a movement as regular and uniform as possible. Then, there is the presence of an ascending and descending movement, through the combined action of the stomach and chest, producing an upward (respectively, downward) volumetric and energetic displacement during the inspiratory (respectively, expiratory) phase. Last but not least, there is also the relevance, for a correct application of the method, of the exchange points between inhalation and exhalation and between exhalation and inhalation, in which it is necessary to produce a sort of energetic shock.

Also in the application of the VELO, many practitioners (the author included) perceive the moment of reversal of the direction of propagation of the energies (at the two extreme points of their longitudinal course) as a critical aspect of the technique, on which it can be advantageous to bring a special attention; see also the discussion in (Trivellato, 2008).

It can therefore be assumed that the instantaneous reversal of the energies could be one of the mechanisms, though obviously not the only one, through which the somatic and energosomatic energies are activated in the execution of the CB and VELO techniques. This assumption seems to find confirmation in an ultra-simplified mechanical model that we will illustrate in the following appendix.

We conclude this section by mentioning, for completeness, that there are other somatic techniques, in addition to pranayama, which can indirectly stimulate the energosomatic energies and promote their partial activation. As an example, we can cite the possibility of working, through specific rubbings and circular movements, on the stiff joints of the physical body, whose hardness is often related to specific emotional energy blockages. An intense bodywork, able to deeply loosen one's joints, can certainly reactivate, to some extent, the energosomatic circulation, thereby promoting (as a result of putting into motion dormant energies) possible vibrational states. Another possibility is the use of sound (*Nada Yoga*), through the

---

[12] The term "circular breathing" is also used in relation to a particular breathing technique employed to play certain wind instruments (such as the Australian didgeridoo), where there is no interruption in the air flow blown into the instrument.

listening and vocalization of specific vibrations, such as those produced by the intonation of certain vowels.

Appendix

On this planet, at the present state of our research, there is an undeniable lack of scientific models able to describe and explain the functioning of our para-anatomy and paraphysiology (and, consequently, able to make falsifiable predictions). For the time being, we only dispose of elementary heuristic models and possible extrapolations from other scientific disciplines, such as physics (Sassoli de Bianchi, 2009).

For example, Alegretti compares the VELO to the functioning of a laser (Alegretti, 2008), but then admits that there are no explanations regarding the origin of the pumping and population inversion mechanisms that would be at the basis of its working. In other words, so far, we can only propose generic analogies, but are not yet able to answer the following simple question: *Why the VELO technique works?*

Not knowing how to answer this question, it becomes even harder to imagine how to improve this technique, to make it even more effective.[13] In this paper, we have proposed to potentiate the VELO by integrating and combining it with a specific Yoga's pranayama, known as the CB, which bears many elements of similarity with it. As already stated, in both of these methodologies the sudden reversal of the direction of the energetic and respiratory motions appears to play an important role, and a better understanding of this mechanism may suggest ways of upgrading this technique. But for this, further investigations, both at the theoretical and practical levels, are required.

With regard to the theoretical aspect, in this appendix we will describe a purely mechanical, ultra-simplified model for the VELO, suggesting a possible explanation for the presumed role played by

---

[13] The two recent works of Trivellato (2008) and Alegretti (2008) can be considered important steps towards a better understanding the functioning of the VELO technique.

the U-turn reversal points of the energetic movement, in the application of the technique.

The consciential shaker model

The first simplifying hypothesis of our model is that the energosomatic substance set in motion by the action of the consciousness is made up of specific particles, which we will call *energons*. We assume that the energons are structured entities, which in addition to their translational degrees of freedom also possess internal degrees of freedom, enabling them to vibrate at different frequencies.

To further simplify the discussion, we assume that an energon can only be in two different states: a ground-state, characterized by a vibrational energy E_0 and a frequency of vibration f_0, and an excited state, of energy $E_1 > E_0$ and frequency $f_1 > f_0$.

We also assume that the substance forming the energosomatic fluid (EF) is very cold,[14] in agreement with the well-known parasensations of extreme low temperatures associated with effects like ectoplasmy, telekinesis, dematerialisation, cold winds, and other phenomena at the interface between the physical and extraphysical.

A further simplifying assumption is that the energons composing the EF move independently of each other (in other words, we neglect in this description the mutual interactions between energons). This means that we are comparing the EF to a sort of ideal gas.

According to the above simplifying hypothesis, it is easy to relate the most typical energosomatic paraperceptions to the state of the EF. For example, the *overall perceived intensity of energies* corresponds to the total energy E_{tot} of the EF, given by the sum of the individual energies of the energons (being them independent, by hypothesis). Moreover, since the kinetic energy of the EF's components is assumed to be negligible (the fluid is very cold), only the internal energy E_{int} of vibration of the energons will significantly contribute to the calculation of E_{tot}. Therefore, if N_0 is the total number of energons in the ground-state of energy E_0, and N_1 the total number of energons in the excited state of energy E_1, we simply have:

$$E_{tot} = N_0 E_0 + N_1 E_1$$

---

[14] We recall that temperature is a measure of the kinetic part of the internal energy of a substance, associated with the movement of the particles composing it.

Obviously, since $E_1 > E_0$, the greater the number of exited energons and the higher will be the total amount of circulating energy perceived by the practitioner.

Another important paraperception is the *overall perceived frequency of vibration*. Considering that the EF is a mixture of energons of two different states, vibrating at frequencies f_0 and f_1, respectively, it is natural to associate the perceived overall vibration to the average frequency $\langle f \rangle$ of vibration of the energons in EF, given by weighted sum:

$$\langle f \rangle = \frac{1}{N}(N_0 f_0 + N_1 f_1)$$

where $N = N_0 + N_1$ is the total number of energons and we have

$$f_0 \leq \langle f \rangle \leq f_1$$

The underlying assumption behind this model is that in our usual intraphysical state we have a maximum of energons in the ground-state (and therefore a minimum of energons in the excited state); but when we apply the VELO (or VESELO) technique, we excite a growing number of energons, thus increasing both E_{tot} and $\langle f \rangle$.

As regards the consciousness-energons interaction, we can hypothesize two modalities: *direct* and *indirect*. The direct interaction would correspond to the ability of the consciousness to directly excite the energons, i.e., to act directly on their internal degrees of freedom. This seems to be the prerogative of a few experienced practitioners, able to activate the EF upon simple mental command, in a few seconds.

The second modality, which is the one that concerns us here, uses instead the ability of the consciousness to act on the external (translational) degrees of freedom of the energons, by putting them into motion, as well as on the ability to confine their movement within a predetermined volume, essentially corresponding to the volume of a cylinder containing the soma of the practitioner. In other words, the action of the consciousness in the execution of the VELO would be twofold:

1. *Setting in motion* of the EF along the longitudinal axis;
2. *Confining* the motion of the EF within a predetermined cylindrical volume.

The reader can for instance imagine a single energon that, under the steady impulse of the consciousness, moves at constant velocity $v = v_0$ along the longitudinal axis of the confining cylinder,[15] and once it reaches one end of it, it bounces back on the confining wall, reversing its direction and acquiring an opposite velocity $v = -v_0$ (see Figure 1).

We are assuming here that the consciousness is able to ideally maintain the *cylindrical confinement field* perfectly stable, without letting it recoil, so that the rebounds of the energon on the cylinder's walls take place in a perfectly *elastic* manner, and therefore the energon's kinetic energy is conserved during its oscillatory sweeping movement (for a given frequency of oscillation). This is what ideally would happen in a standard correct application of the VELO technique, when the practitioner can reverse the direction of EF without producing undue decelerations and delays.

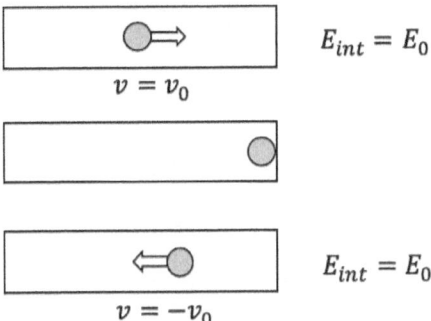

$$E_{int} = E_0$$
$$v = v_0$$

$$E_{int} = E_0$$
$$v = -v_0$$

Figure 1 A hypothetical energon bouncing back in a perfectly elastic way on a static confinement field. The energon's internal energy E_{int} is the same before and after the collision with the confining wall.

Let us try now to translate into the language of our simplified model the suggestions of the *Bhagavad Gita*, i.e., "to offer one movement into the other," or of the *Vigyana Bhairava Tantra*, i.e., "to produce a collision of the two movements." Imagine that a moment before the energon reaches the end of the confinement field the inversion of its movement is anticipated by the field itself. In other words,

---

[15] More precisely, because the EF has its own viscosity, the consciousness must apply a constant force on the energon, to compensate for its resistance to flow. When this happens, the sum of external forces is zero and the energon can move at constant speed (like a free particle).

suppose that a moment before the energon arrives, say, at its right boundary, the whole field (or possibly just its right boundary) produces, under the impulse of the consciousness, a rapid anticipatory movement to the left. In this case, the rebound of the energon would not be any more elastic, and after the collision it would have acquired a surplus of energy (see Figure 2).

This surplus could then produce two effects: (1) increase the kinetic energy of the energon, or (2) increase its internal vibrational energy E_{int}. In other words, we can assume that, if the practitioner is able to impress a precise and sufficiently intense impulse at the extreme points of the oscillating movement, by anticipating the motion reversal of the energons, then at each change of direction it may be possible to excite a certain number of them, thus increasing the average vibrational frequency of the EF.

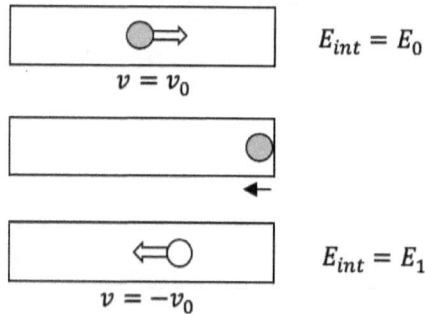

$$E_{int} = E_0$$
$$v = v_0$$

$$E_{int} = E_1$$
$$v = -v_0$$

Figure 2 A hypothetical energon acquiring a surplus of internal energy, when bouncing back in a non-elastic way, because of the anticipatory axion the confinement field.

Certainly, being the present model extremely naïf, its heuristic could be totally wrong. It is also clear that it can only highlight one of the many aspects of the complex activation mechanism subtended by the VELO technique.

In fact, as is clear to any sufficiently experienced practitioner, the translational movement of the EF through the complex web of energosomatic channels and chakra, is for itself already able to produce, regardless of the reversal mechanism, a significant level of activation of the same, and of course this effect cannot be explained in the ambit of the oversimplifications of the model we have presented here.[16]

Bibliography

Alegretti, Wagner (2008). An Approach to the Research of the Vibrational State through the Study of Brain Activity, Journal of Consciousness, Vol. 11, No. 42, p. 217.

Monroe, Robert A. (1977). *Journeys out of the body*, Broadway Books, New York.

Patanjali (2003). *Aforismi dello Yoga (Yogasutra)*, a cura di Paolo Magnone, Promolibri Magnanelli, Torino.

Sassoli de Bianchi, Massimiliano (2009). *Interdimensional Energy Transfer: a Simple Mass Model*, Journal of Consciousness, Volume 11, No. 43, p. 297. (Also published in this volume).

Sassoli de Bianchi, Massimiliano (2018). *Sàdhàna*, in: AutoRicerca, Volume 15.

Trivellato, Nanci (2008). Measurable Attributes of the Vibrational State Technique, Journal of Consciousness, Vol. 11, No. 42, p. 165. (Also published in this volume).

Vieira, Waldo (1994). *700 Experimentos da Conscienciologia*, Rio de Janeiro, RJ, Instituto Internacional de Projeciologia e Conscienciologia.

Vieira, Waldo (2002). *Projectiology, A Panorama of Experiences of the Consciousness outside the Human Body*, Rio de Janeiro, RJ – Brazil, International Institute of Projectiology and Conscientiology.

Vieira, Waldo (2003). *Homo Sapiens Reurbanisatus*, Foz do Iguaçu – Paraná – Brasil, Associação International do Centro de Altos Estudios da Conscienciologia – CEAEC.

Note: This article was previously published in: Journal of Consciousness 12, 2010, pp. 283-316, with the title: "From Yoga's pranayama to Conscientiological VELO: a proposal for an integrative technique." An Italian translation is also available in: *Lo Stato Vibrazionale*, AutoRicerca, Numero 1, Anno 2011, pp. 101-137.

---

[16] In more general terms, we can assume that the multiple collisions experienced by the energons travelling inside the complex energosomatic structure are in part responsible for their excitation. This would be the case because, every time an energon collides with another energon, or with an element of the energosomatic structure, there would be a non-null probability that part of its kinetic energy is converted into internal energy, thus increasing and the total internal energy and the average frequency of the EF. Accordingly, the consciousness will have to apply an additional force on the energons, to compensate for the corresponding loss of kinetic energy, in order to preserve the overall speed of the longitudinal impulse.

AutoRicerca

Bioenergy and vibrational state detection via fMRI: preliminary results and analyses

Wagner Alegretti

Issue 20
Year 2020
Pages 123-153

Abstract

Bioenergy (prana, chi, orgone or subtle energy) has been known for millennia but has eluded science in respect to its detection, measurement and theoretical understanding. To help cover this serious gap in the comprehension of reality and consciousness manifestation, the author designed and conducted three series of experiments using fMRI (the latest in December 2014) focused on the research and understanding of: 1. neurological changes of the brain during execution of bioenergy techniques and vibrational state; and 2. effects of bioenergy over matter and the mechanism of consciousness-matter interaction via bioenergy. This paper intends to present thought-provoking findings and their preliminary analysis, mainly of bioenergy transmission to different types of substances, like a "fMRI reference phantom", while discussing some of their consequences and possible uses, including the possible future development of a "bioenergy technology". It aims mostly to motivate researchers to replicate and discuss his experiments. Future experimental approaches will also be proposed and discussed.

AutoRicerca, Issue 20, Year 2020, Pages 123-153

1 Objectives

The main objectives of the present paper are:

o Communicate some thought-provoking findings with bioenergy effects detected via fMRI
o Present preliminary analysis of these findings
o Propose a methodology for the research of the vibrational state and bioenergy
o Stimulate other researchers to replicate these experiments to confirm (or to falsify) those findings
o Collect suggestions and constructive criticism for improving the next series of experiments
o Propose further experimental approaches
o Lay the foundations for a Bioenergy Technology
o Establish hypotheses and initial theories
o Discuss possible future applications for this line of study.

2 Introduction

Since ancient times humanity mentions a form of subtle energy, parallel to the materiality of the day-to-day, but not so far from human beings that would not be felt and known by people more sensitive or those aware of their existence.

Having received several names over the centuries, such as prana (Yoga), chi / qi (Chinese medicine), orgone (Reich), magnetic fluid (Mesmer), vital fluid (Kardec), life force (Hahnemann), astral light (Blavatsky) and etheric force (dowsing), bioenergy, as it is called in this paper has eluded modern science not only with regard to its detection and measurement but also in respect to its theoretical conceptualization and modelization.

There have always been stories, accounts and reports of people claiming to be able to feel or see this form of subtle energy in living things, the environment and other people and sometimes also be able to apply it to improve their own quality of life or of others around them, including in situations such as spiritual healing.

Studies have been carried out in various areas of human knowledge, including Parapsychology, alternative medicine and biology (morphic resonance, biophotons) in order to clarify the nature of this energy. But, why in despite of millennia of cases and of these research works and others from some pioneers like Wilhelm Reich, Semyon Kirlian and Konstantin Korotkov, the reality of bioenergy has not been accepted by the scientific community and general public? Perhaps the scientific quality of the evidences and experiments have not being good enough.

To better understand this form of energy as well as its role in the expression or manifestation of consciousness, the author has been conducting various experiments since 1984 in order to detect and measure this bioenergy. While still studying electronic engineering in the university, the author used various instruments and components available at the time, such as magnetometer, Geiger counter, semiconductors, and others, but without relevant or reliable results.

Based on the assumption that living things, including humans, are natural transducers of bioenergy, since physical life would be itself the biggest manifestation of bioenergy, and also knowing about the ability of many people, throughout the centuries, to feel bioenergy, the author took off from the hypothesis that the answer to his quest for an bioenergy detector principle was in organic matter.

To avoid the use of living things, both due to the ethical problem with that and the unreliability of the results due to interference of the natural dynamics of the internal metabolism and even to behavioral changes, organic substances "in vitro" were chosen. So, proteins were picked for being universal in the central processes of life and also due to their flexibility in reactions. Indeed, their behavior depends not only on their chemical formula but also on their particular 3D folding or shape.

Thus, transducer prototypes were built based on an aqueous gel of collagen, of which its electrical resistance was measured. Some promising results were obtained with this principle but unfortunately the author had to put this line of research on hold for reasons whose explanation would go beyond the scope of the present writing. However, he intends to return to this line of experimental

126

research as soon as possible.

With the goal of establishing a new line of investigation allowing for the identification of parameters for the development of a methodology of bioenergy research, this author presented in 1990, during the 1st International Congress of Projectiology and Conscientiology (Rio de Janeiro, Brazil), a conference entitled "Bioenergy Technology", where he presented the results of his research done from 1984-1988 (Alegretti, 1990). In that ambit, the principles of 'bioenergy technology' were also presented, as well the initial experimental results, the discussions about its relevance and applications and also the planning of the phases for its development.

However, over time, the author tried to stay up-to-date of the progress of science, scientific instrumentation and research in medicine and neurology. A resource exploited in medical analysis always called his attention: the nuclear magnetic resonance imaging (MRI), and mainly a variation of it that allows the monitoring of the functioning of the brain during specific tasks: the functional magnetic resonance imaging (fMRI), which is based on the BOLD (Blood Oxygenation Level Dependent) technique to register the change of magnetic resonance of the hemoglobin when passing from the oxygenated state (arterial blood) to deoxygenated (venous blood). This technique has been used in several scientific researches on brain function and also is now being used in clinical exams, especially as a way to get previous data before certain brain surgeries.

So, he had the idea of using fMRI to monitor what happens in the brain when someone consciously and willfully controls his bioenergy to provoke an internal intensification of one's natural energetic vibration, a phenomenon known as *vibrational state* (VS) (Trivellato, 2008). The VS has the advantages not only of creating an unusual intensification of the personal bioenergy field, but also of being producible by the will through a standardized technique: VELO (voluntary energy longitudinal oscillation), which can be learned, developed and applied by anyone. At that time, he published a paper in the Journal of Consciousness[1] (Alegretti, 2008) proposing this particular research and discussing its methodology, possible benefits and applications. But as it will be explained below this line of research expanded to also understand what

---

[1] Formerly known as: Journal of Conscientiology.

happens in the brain during the intentional absorption and transmission of bioenergy, including to different media, like fMRI phantoms, a potato and an egg.

The possible success of this approach may open doors for future similar studies, by showing that the research of certain consciential phenomena, until now considered subjective or beyond the scope of intraphysical analysis, is doable and viable through conventional third-person methods and techniques. Besides proving this methodological line, such demonstration of executability and viability would certainly stimulate studies of other consciential and parapsychic (not purely physical) phenomena which are even more complex.

As bioenergy is claimed by many to be a central aspect of consciousness manifestation, not only through the expression of biological life itself but mainly in a class of phenomena called paranormal, non-physical, extra-physical, psychic or parapsychic, its detection and highlight could be half way to a more solid evidence for the objective nature of consciousness, as understood in the so-called consciential paradigm (Vieira, 2002). According to this paradigm, the bioenergy would be the bridge between the nonphysical consciousness (independent from the body) and the physical world of the ordinary forms of matter-energy (hard problem of consciousness).

The data and preliminary qualitative results from the 3 series of experimental sessions already conducted by the author, as presented below, seem to point towards the validity of the hypotheses and viability of this research and of the methodology adopted.

In parallel to the bioenergy detection line of research, as explained, the author also conducted some pilot experiments with electroencephalography (EEG) to search for neural correlates of some consciential states and bioenergy procedures. In 1991 he had the first opportunity of doing some personal experiments with the lucid projection (out-of-body experiences) while monitored by an EEG (and other devices for measurement of several physiological parameters) in a sleep study laboratory of a hospital in the city of Porto Alegre, Brazil. On this occasion, the author also installed the VS voluntarily, as to allow for the observation of the changes in brain waves pattern that could be generated as a result of this process.

Also, in December 2007, another series of experimental sessions of were done, having this author and Nanci Trivellato as objects of study, at a neuroscience laboratory in the city of Natal, Rio Grande do Norte, Brazil. Those experiments focused on registering brain activity through digital EEG during the production of VSs and partial projections (Alegretti, 2008).

Continuing this specific line of investigation, a pilot study with a more advanced EEG equipment and a better scientific support has just being initiated in partnership with the TransTech Lab at the Sofia University in Palo Alto, California, USA.

Another precious source of information for this line of study has been the research done through the bioenergy evaluations of individuals, mainly of students of the course 'Goal: Intrusionlesness', given by this author in partnership with Nanci Trivellato since 2002, at the International Academy of Consciousness (IAC). Those individual evaluations brought up observations, findings and further questionings about the VS and bioenergy procedures, generating new hypothesis and allowing for the perfecting of the experimental protocol of the research discussed in this paper (Trivellato & Alegretti, 2005).

Some of the fMRI preliminary results and analyses have already been shared with fellow researchers in this field such as Dean Radin (IONS – Institute of Noetic Sciences), Beverly Rubik and Harry Jabs (IFS – Institute for Frontier Science), Ivan Lima (North Dakota University) and members of the FMBR – Foundation for Mind-Being Research, in Palo Alto, California, USA.

The EEG line of research recently inspired Rute Pinheiro (UFRN – Universidade Federal do Rio Grande do Norte, Brazil) to initiate and conduct a similar investigation (Pinheiro, 2013).

3 Hypotheses

Assuming the Consciential Paradigm (that consciousness is objective, multidimensional, a property of the universe, neither reducible to matter-energy, nor a mere product or epiphenomenon of it) this research is founded on three basic hypotheses:

1. Bioenergy is real and objective, being able to interact with matter and other form of energies, as in the manifestation of life;

2. The VS is an objective condition, not being just imagination, illusion or sensory hallucination of the practitioner;

3. The VS and other bioenergy procedures and regimens are, in certain circumstances, accompanied by detectable changes in the human brain or can cause alterations in such (some temporary and, perhaps, others more permanent).

Based on the specific knowledge existent today about the VS, which is still relatively limited due to the lack of systematic research about this phenomenon conducted until this moment, it is not yet known whether there are true VSs, of high intensity, that do not produce any level of repercussion on the soma, or even if all the intense VSs will cause a repercussion on the soma.

Within the numerous types of repercussions of the VS so far observed in first-person experiments, it is reasonable to assume that some VSs will produce a bigger effect on the soma, while others will concentrate its effects more directly on the energossoma, or maybe, on the vehicles more subtle than this one. This way, it is anticipated that there would be VSs that do not produce any somatic effect that can be registered by technical apparatus of physiological or neurological monitoring existent nowadays (or, more probably, that produce a too weak somatic effect to be detected).

So, this research initiated on studying the VSs and bioenergy procedures which effects can directly reach the physical body, while recognizing that the VS happens primarily at the level of the energosoma, and that this subtle energy body acts as an interface between the consciousness (or better, between the more subtle vehicles of manifestation of the consciousness) and the physical body.

Another aspect that suggests that the occurrences of somatic repercussion of the VS are a common condition is the observation (through personal experience of this author and other people's reports and publications (Trivellato, 2014)) that the great majority of the VSs experienced by the consciousness when in coincidence with the physical body are felt also in the body, or at least as physical sensations (for the less sensitive individuals towards subtle energies, those would anyway be felt mainly on and by the physical body). This observation points to the logical assumption that, since the sensations and – or at least some – energosomatic effects manifest frequently (and at times intensively) on the soma, the VS probably also produces changes on the soma that can then be measured.

4 Objectives of the research

The experimental development of studies in this area has sufficient merit, since it would allow for promoting the following possible relevant results and discoveries, among others. So, within a broad view, in short, mid and long terms, the objectives of this line of research are:

○ Create means for the detection of bioenergy;
○ Measure such bioenergy;
○ Gather a sufficient number of different and converging evidences that support the theory of objectivity of the bioenergy (see hypothesis 1);
○ Investigate its characteristics and properties and establish the general laws that govern it (analogy: history of the study of electricity and magnetism up to the Maxwell equations) to create a theoretical framework (able to predict findings) for understanding of bioenergy within a multidimensional view of the nature of consciousness;
○ Understand the biological and, more specifically, neurological effects of bioenergetic processes, like the VS, and bioenergy in general;
○ Identification, categorization and description of the neurological effects provoked by the VS or concomitant to it (see hypothesis 3);
○ Demonstration of the VS as a real and objective energetic phenomenon (see hypothesis 2);
○ Characterize the VS as a stand-alone state, different from other neurological, physiological or mental states;
○ Collection of data and findings for a better comprehension of the VS itself;
○ Better understanding of the processes and factors involved in

the development and effective installation of the VS, allowing for the generation of more effective didactical methods and more exact descriptions, capable of promoting better energetic self-control for the community of practitioners of the VS technique;

o Classification of the VS according to the level of effect on the soma, and then according to the types and intensity of the repercussion on the energosoma and other vehicles of manifestation;

o Clarify the mechanisms of the interfaces psychosoma-energosoma-soma and parabrain-brain (hard problem of consciousness), so understanding how the consciousness (a non-physical entity) interacts with the physical dimension, or, in other words, how it can control and sense the physical body and universe;

o Search for and development of new therapeutic applications of the VS in particular and bioenergy in general;

o Build a bridge between the study of consciousness under a multidimensional paradigm and the more specific neuroscientific approaches, enriching both sides.

5 fMRI experiments

The somatic and inter-vehicular repercussions of the VS and other bioenergy procedures and regimens can in principle be studied according to different criteria and techniques. The ideal would be through a bioenergy technology that would allow for the direct detection and measurement of the energies. However, this technology does not yet exist. With regard to indirect methods, due to its advanced design and the fact of being based on a pure quantum effect (the alignment of nuclear spins in a magnetic field) the fMRI seemed to be a good choice for initial tests.

In other terms, to allow for the realization of a more objective exam of the effects of the VS and the systematic comparison of the results, the most effective and consistent method seemed to be the detection of the neurophysiological alterations of the practitioner through the best technology for analysis of neurological functions available today. This is also because the method allows for the easy replicability of the experiments by any researcher (even those with a more skeptical approach or who has never felt or produced a VS).

So, the fMRI experiments were performed by the author in 3 series, as listed below, using MRI systems graciously lent by generous clinics and radiologists in Brazil, which allow their use over weekends and other free periods of time when the systems were not being used to perform clinical exams. Also, the chief radiologist of each clinic took part on most or some of the experiments of each series, monitoring the results and contributing with suggestions and advices. Furthermore, MRI technicians were made available for the running of the tests.

o   Series 1: 2009-December – 1T Philips machine (at the Clínica Radiológica de Anápolis, with Dr. Paulo Eduardo de Jesus);
o   Series 2: 2010-March – 1T Philips machine (at the Clínica Radiológica de Anápolis, with Dr. Paulo Eduardo de Jesus);
o   Series 3: 2014-December – 3T Philips machine (at Ultramed Clinic, in Londrina, with Dr. Fábio Takeda).

As explained before, these experiments started with the initial objective of studying neurological changes of the brain during the VELO procedure and consequent VS. Nonetheless, as it will be better explained later it widened to encompass also the effects of bioenergy over matter and the mechanism of consciousness-matter (so far observed on water solutions, potato and egg) interaction via bioenergy.

Basic protocol

Each individual experiment (data acquisition session of the fMRI system) was planned to be divided in 3 periods, in the following way:

o   1st period: initial resting (inaction) with the subject or "energizer" just being as relaxed as possible, but conscious, to establish a baseline or reference state;
o   2nd period: action (which was different for each group of experiments or sessions);
o   3rd period: post resting (inaction), with the subject or "energizer" just being as relaxed as possible again, but conscious, to establish an afterglow or a second reference state but also to

test any possible lingering or delayed effect of the previous action period.

The average duration of the individual experiments was of 3 minutes, divided equally in 3 periods of 1 minute.

As it is the case in all fMRI tests, at the beginning of each series of experiments with a subject, being a person or object, a regular MRI test for static, anatomical data acquisition, was run, which would take between 40 to 50 minutes. Later, the fMRI (BOLD) tests were run and the images imposed over the static or "anatomical" images.

In the cases when people were being the subjects, after the static MRI imaging, a standard "finger tapping" test was run to verify the proper functioning of the system, adjusting the parameters to avoid a too low sensitivity (that would cause no images during the action phase – in this case just the "finger tapping") or a too high sensitivity (that would pick noise or artifacts images during the inaction periods).

As in many experiments of this type, an intercom between the experiment (MRI) room and the control room where the researchers and the computers were was used to guide the subject actions and to facilitate and enrich many of the results, for allowing, among other possibilities, that the subject announces to the researchers beforehand of what he wants to do, or what according to him just happened or, yet, what is taking place at that exact moment.

The procedure for the attainment of the VS (through the VELO technique) presents a certain complexity in terms of the several mental commands applied and their synchrony, when compared to other simpler actions such as moving a finger or seeing a colored light. For this reason, when it is taken into consideration also all the procedural vices and bad habits used by many practitioners of the VS, the necessity of establishing a protocol with strict criteria that allows for the isolation, and later removal, of those interferences from the 'pure' VS becomes important. Otherwise, such factors could generate spurious effects in the results of the research.

The recording and analysis of the brain activity in certain stages described below (apparently disconnected of the objective of the bioenergy experiments) have the purpose of working as a 'control' reference for the analyses, since they will allow the comparison of the results with the data obtained from the voluntary and correct application of the VELO and possible installation of the VS with great intensity.

So, as previously proposed in by the author (Alegretti, 2008), for phase 1 of the experimental protocols, during the "action period" the following actions were performed:

1. Relaxation/rest only (so, resulting in continuous inaction during the 3 periods);
2. Conscious and voluntary rhythmic inhaling and exhaling only (as some people do while trying, erroneously, to perform VELO) - slow in the beginning and then with its gradual acceleration;
3. Conscious and rhythmic moving of the eyes up and down, with the eyelids closed (as some people do while trying, erroneously, to perform VELO) - slow in the beginning and then with its gradual acceleration;
4. Visualizing or imagining the rhythmic movement of bioenergy up and down, slow in the beginning and then with its gradual acceleration, simulating VELO, without real intent to move energy;
5. Scanning of the perception of the sensations of the body, up and down. In other words, only sweeping of attention and perceptive focus through the soma - slow in the beginning and then with its gradual acceleration. In this case, the subject tries to concentrate exclusively on the existence of the part of the body that is being focused on. Such focus will move smoothly, continually and cyclically up and down along the body (from feet to head and vice-versa);
6. Effective installation of the VS through the correct and vigorous application of the VELO.

In accordance with the above discussion, the protocol here presented can be used with any data collection resource, like fMRI or EEG, or even, in the future, with a direct bioenergy detection system.

Even though the procedure described on Stage 1 is unnecessary for the production of the VS (though not counterproductive), its recording and study are essential to establish the baseline, meaning, the specific basic resting neurological condition specific and particular to each participant. This resting condition will be an important reference for posterior comparisons and analyses.

The six experimental steps here described were planned to make possible, in the data analysis, the basic strategy of subtracting from the data set relative to Stage 6 the signals obtained in the previous stages, determining in this way the profile of the VS itself, separating its neurologic 'signature' from the other associated neurological processes, be them natural or derived from the technique application (assuming here that the neural structures supporting cognitive and behavioral processes combine according to a simple additive logic).

The steps described in the Stages 2 to 5 aim at simulating a pseudo-execution or erroneous execution of the technique of installation of the VS, having been inserted to take into consideration also the habits (some of them inappropriate) common in the application of the VELO technique. Evidently, the data obtained during those stages have the objective of being more than just 'noise to be removed', since the careful analyses of those can lead us to a better understanding of the mechanisms of the technique of the VELO and of the pros and cons factors to the attainment of the VS. Furthermore, they will allow for a clearer verification of the influence or not of those somatic or mental 'crutches' over the execution of the VELO and the VS.

6 Description of some of the results

With VELO and VS

For the actions listed above, during 2 through 5 there was no relevant difference when compared to the "action" 1 (whole session of resting) or the inaction periods before and after each of those respective actions (there was very little or no BOLD activation).

VELO produced significant BOLD activation and when VS occurred the activation was comparatively stronger.

During VELO, and even more during VS, there was intense activation of many and different areas of the brain (see Figures 1 and 2), distinct from and perhaps even stronger than that of normal actions or tasks, but certainly stronger when compared to the previous finger tapping (see Figure 3).

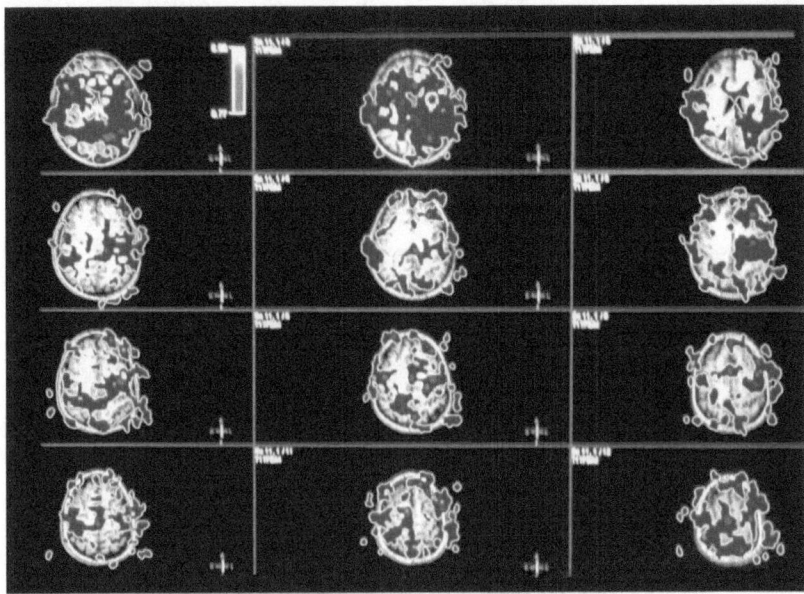

Figure 1 The fMRI images of the brain during the VS, with an intense and generalized BOLD activation (series 1).

Up to this moment it was not observed any specificity or pattern of active areas of the brain or cerebellum, for the same subject or even across different ones. For the VELO experiments, the subjects were: *Nanci Trivellato, Luis Cláudio Gonçalves*, and the author.

During VELO and VS, images (BOLD-like signal) appeared outside of the skull region, something that in principle should not happen. The first explanation was obviously of artifacts. Nonetheless these extra-cranial signals remained consistent throughout most of the VELO-VS sessions during series 1 and 2, but not during series 3. It is important to remember that the coil room (exam room, where the subject stays during the tests) is heavily shielded to avoid interferences, inwards or outwards. Even the possibility of a shower of cosmic rays was considered, but it was discarded when the same effect repeated many times.

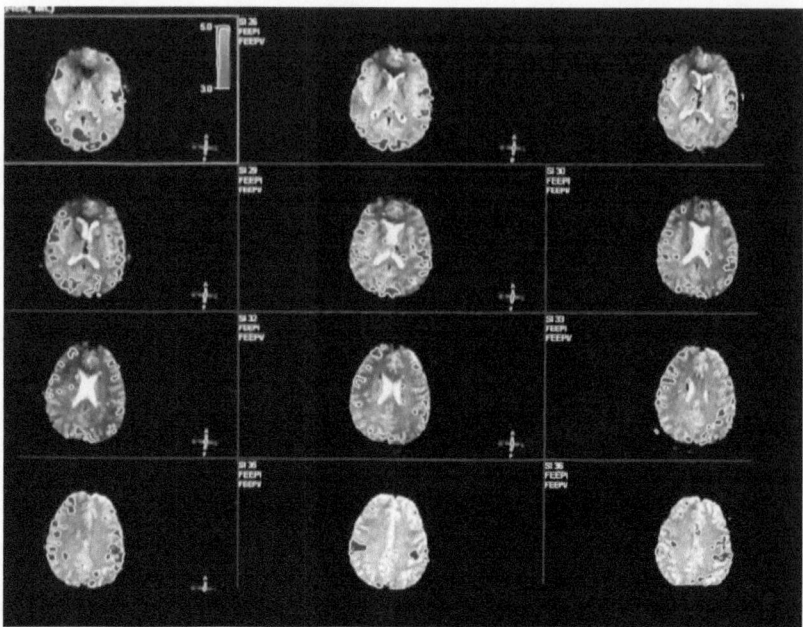

Figure 2 The fMRI images of the brain during the VELO, with a widespread BOLD activation (series 3).

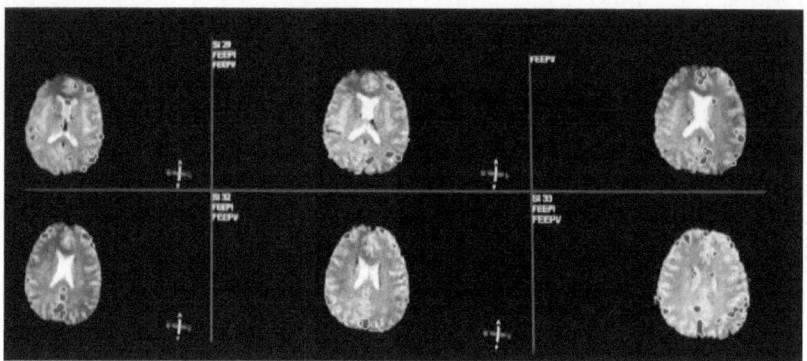

Figure 3 The fMRI images of the brain during the execution of the "finger tapping" (series 3).

It was considered that small head movements could have provoked the out-of-the-head images. During the series 1 and 2 the radiologist said he reviewed the data and confirmed there was no movement. Even so, extra experiments where conducted to try to control for that and the subject, during the action period, produced small amplitude and fast trembling of the head by contracting strongly the neck muscles, even more pronounced than any micro movements possibly produced during VELO and VS. A few and faint BOLD images were produced out of the head, but significantly weaker than with the real VELO and VS.

With externalization of bioenergy

Expanding from the original plan of studying initially only the VELO procedure and VS, due to the unexpected finding of the out-of-the-head signals, it was decided, already at the series 1 of experiments, for the subject to actively exteriorize bioenergy from the head. The result was the production of out-of-the-head images even stronger, across all experiments of this sort (Figure 4).

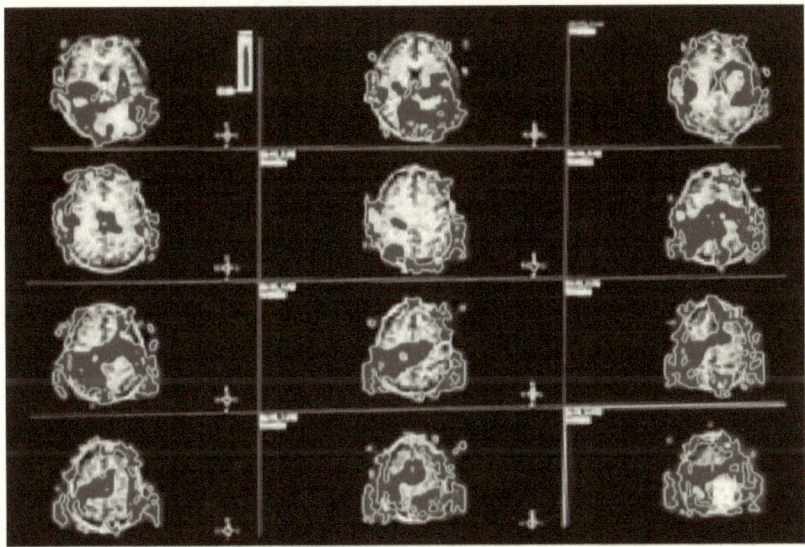

Figure 4 The fMRI images of the brain during the execution of exteriorization of energy from the head (series 1).

To go deeper on this possible important finding, of apparent direct effect of bioenergy over a non-organic (and no living) medium, the author, also the active subject (the "energizer") for this particular series of experiments, had the idea of using a MRI phantom as the "the fMRI or passive subject," inside of the secondary coil (the one normally put around the head of the person when under MRI examination). The phantom used during the series 1 and 2 (Figure 5) was basically a plastic bottle with water containing copper sulfate, sodium chloride, sulfuric acid and *arquad* (a surfactant and preservative agent).

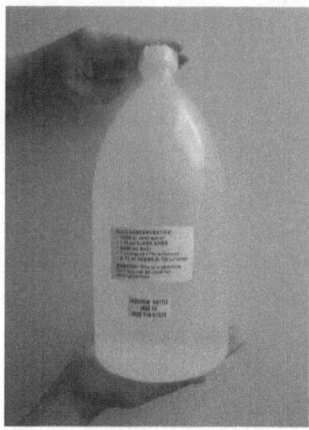

Figure 5 The liquid sample used in the experiments of series 1 and 2.

With the phantom inside of the coil and the energizer lying on the bed with abdomen down, arms stretched above the head with both hands positioned out and at the sides of the secondary head coil, the experiments were run (Figure 6). The hands were kept out to avoid or at least reduce any possible influence of hand movements or blood circulation changes during the "action period" of the experiment. A strong BOLD signal occurred again, during the energization period, which stayed and got slightly stronger even during the 3rd period (Figure 7). This result was confirmed with all of the other repetitions (sessions) of the same type of experiment. During the sessions with no externalization of bioenergy during the "action period", but with exact the same setup, (meaning with hands still at the external side of the secondary coil surrounding the phantom) there was no BOLD signal. So, the mere presence of the hands was not producing BOLD signal and corresponding images.

During the 3rd series of experiments (3T machine) a different phantom (the one available) was used. It was spherical and the solution was of water and chromium chloride. Surprisingly and unexpectedly there was no relevant result. This could be because the head-neck coil was being used (see the discussion in the next section) or due to the specific chemical profile of it.

During the 3rd series of experiments a potato was also used. No relevant results were obtained, just a few very little groups of pixels of BOLD image (Figure 8).

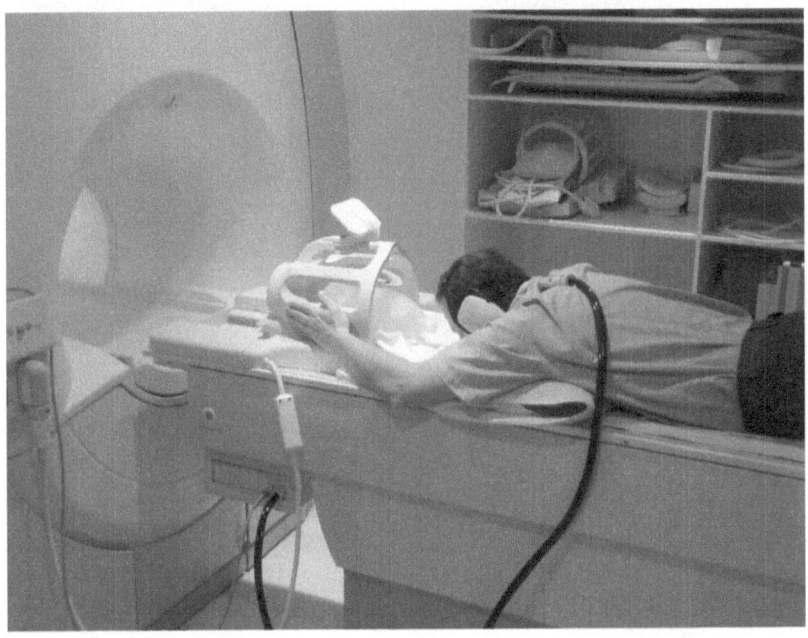

Figure 6 The experimental setup for the transmission of bioenergy to the reference phantom (series 2).

Also, during the 3rd series of experiments an unfertilized chicken egg (off a common type from a supermarket) was further used. The results obtained were the most significant, for being very intense and very synchronic with the periods on and off of the experiments. To get results even more reliable, in this case the hands of the energizer (the author) were kept not at the sides of the head coil, but further down along the bed, about 10 cm far from the base of the coil. During the first inaction-resting period there was no BOLD activation. Then, 4 seconds after the beginning of the energization (transmission of bioenergy to the egg) the BOLD signal starts to appear, and it keeps intensifying up to the moment the energizer receives the instruction to stop. Then, after about 5 seconds the signal starts to fade until it disappears. During the activation the BOLD image revealed an internal structure inside of the egg (not noticed before), mainly inside of the yolk. See Figure 9 (before) and Figure 10 (during).

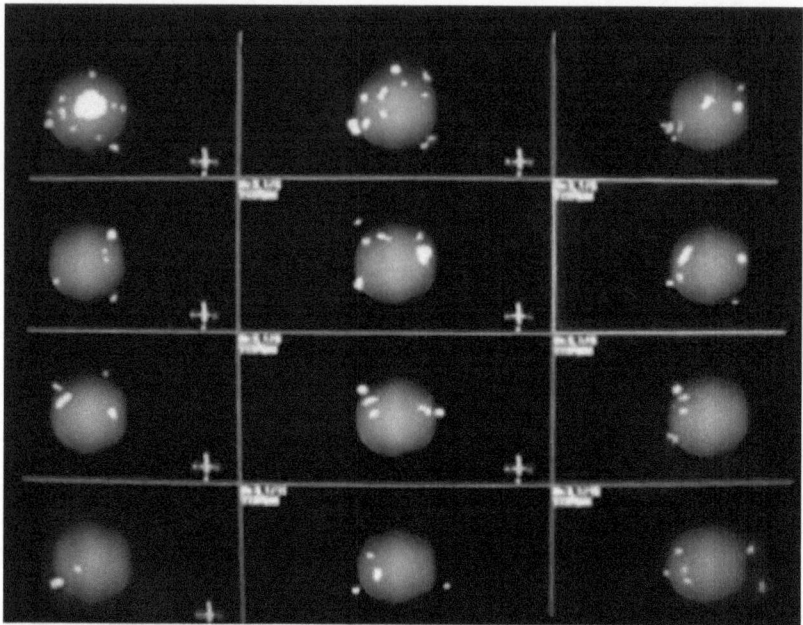

Figure 7 The fMRI images of the reference phantom during the bioenergy transmission by the energizer (series 2).

Successive experiments with the same egg and energization procedure started to show a growing tendency for lingering effects of the bioenergy. In other words, with each experiment of this type there was a stronger and more lasting BOLD activation after the cessation of the exteriorization of bioenergy.

7 Preliminary analysis and discussion of results

The MRI head coil is more sensitive than the head-neck coil and should be the one used in this kind of experiments. The sessions performed with a head-neck coil, used only at the beginning of the 3rd series of experiments, with the 3T system, produced no relevant or reliable results. All experiments done with that coil had to be repeated with the head coil. In spite of knowing that the head-neck coil was somewhat less sensitive, it was used to try to observe a wider area of the body and around of the head.

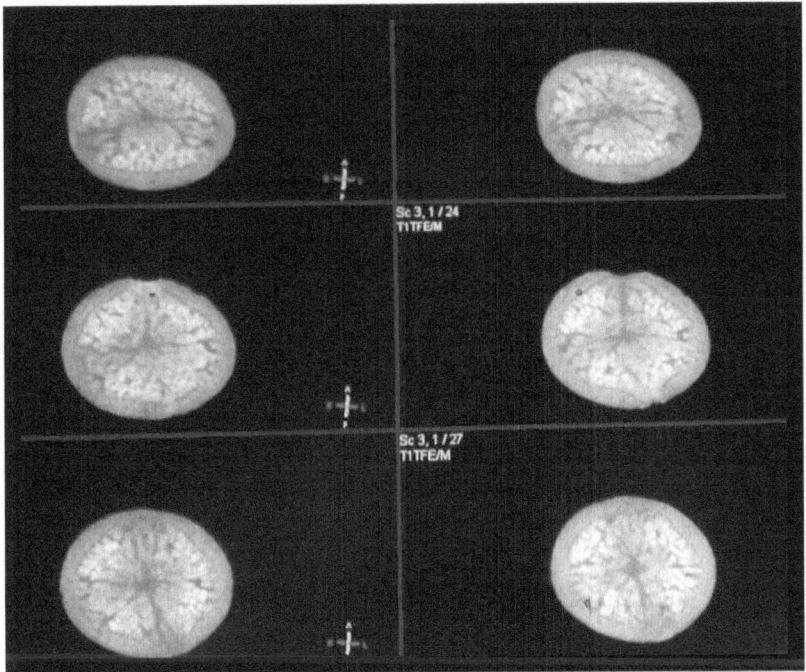

Figure 8 The fMRI images (detail) of the potato during the bioenergy transmission by the energizer.

In respect to the out-of-the skull images during VELO and externalization, could the intensified bioenergy field around the head be altering the MR (magnetic resonance) properties of one or more of the gases of the air (related to their spins or collective behavior of its molecules), to the point of producing a BOLD-like signal that can be picked by a machine able to detect magnetic variations, in a way similar to what happens with the hemoglobin? If so, which gas(es)?

The same effect occurred with the phantom. Was the bioenergy acting equally on all the different types of atoms or molecules of the constituents of the phantom or more/only on one or a few of them? Which ones would be more responsive to the bioenergy? If such, what kind of "artifact" could be produced that way?

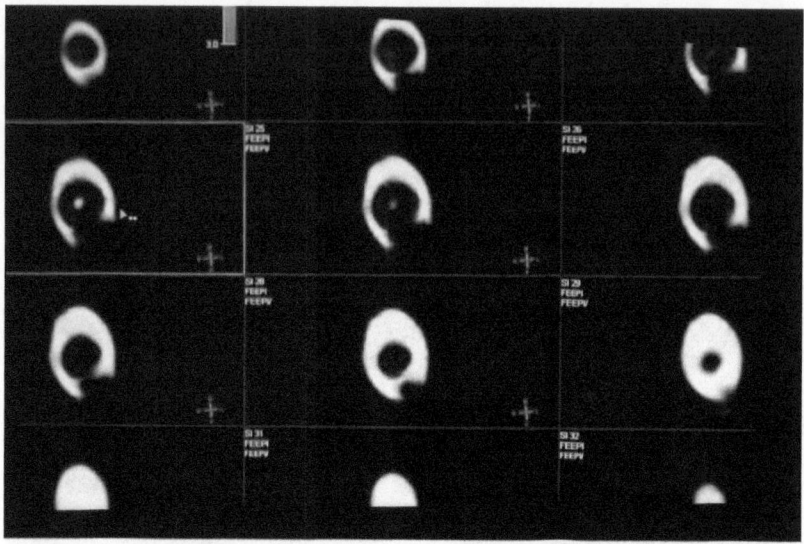

Figure 9 The fMRI images of the egg before the bioenergy transmission by the energizer, where no BOLD signals are detected (series 3).

In the case of the chromium chloride phantom (3rd series of experiments) the absence of BOLD activation could be because the head-neck coil was being used or due to the specific chemical profile of it. Also, if the activation seen in the other experiments was just artifacts (caused for instance by hand movements), why it did not happen in this case?

It is of course necessary to repeat this kind of experiment with many different types of MRI phantoms, preferably always using a coil of same sensitivity, to look for possible patterns.

Why there was such a little result with the potato? There is water and many complex molecules in it. Why none of them reacted significantly to the bioenergy that particular time? Again, if the activation seen in the other experiments was just artifacts (caused for instance by hand movements), why it did not happen in this case? Why there was such a strong effect with the egg? Why, with the bottle phantom, different from the egg, there was no lingering activation after cessation of the energization? It was because of the medium per se or it could be due to a particular difference in the data acquisition, analysis and presentation by the fMRI system? What is the physical mechanism (detectable via the fMRI technique) that could explain these different observations? In that respect, it is interesting to observe that the spin of subatomic entities, which is at the origin of the changes in magnetization observed via the fMRI technique, is a quantum observable having no classical analogue. Could it be then that the spin, with its typical non-spatial properties (Aerts & Sassoli de Bianchi, 2015), could provide a sort of bridge between ordinary matter-energy and the more subtle bioenergy?

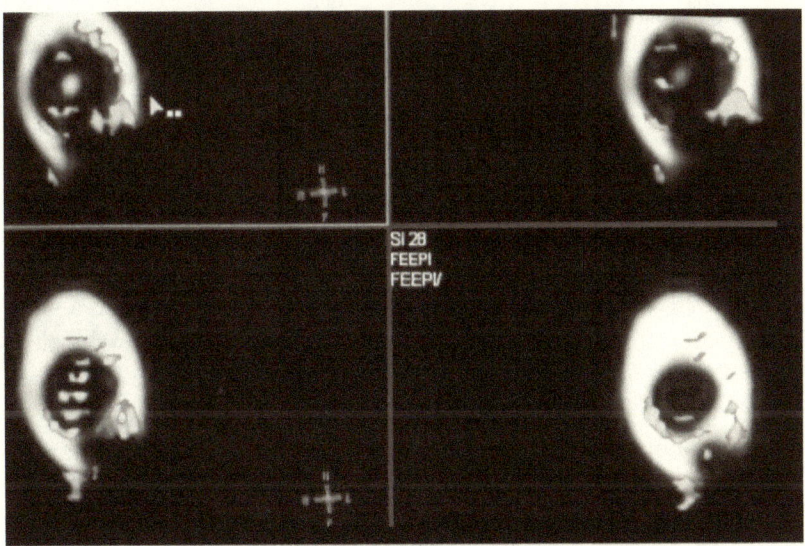

Figure 10 The fMRI images of the egg during the bioenergy transmission by the energizer, with the presence of a strong BOLD signal (series 3).

As for the BOLD image revealing an internal structure inside of the egg, mainly inside of the yolk: Which substances predominate there, that can be easily activated by bioenergy? Perhaps some specific proteins that could be used in further experiments, to provide a more efficacious transducer (as discussed earlier), without the need for the complex fMRI.

If the lingering effect of the bioenergy is confirmed, would it change with different materials? Will it be possible to indicate different bioenergy decay rates, for different materials?

The occurrence of activation (BOLD images) in the air, bottle phantom and egg bring the very important question if the images detected in the brain during VELO and other bioenergy procedures are the result of the mind-brain action of moving the bioenergy or if the brain is being affected by the bioenergy (or a combination of both things)? Can the brain (a physical organ) act on the energy of the more subtle bodies of the consciousness or is the opposite (or a combination of both things)?

As seen with these preliminary experiments, they still mix neurology, biology and physics in a way that is still difficult to separate their respective domains of study. It is necessary to design experiments to isolate them as much as possible.

Under the very strong field of the 3T machine it was very difficult for the subject (the author) to "work" with or to move his bioenergy. Why? Was this due to the effect of a magnetic field 3 times stronger that with the system used during series 1 and 2? Or, could it be due to other yet to be identified physical factors, or also nonphysical ones, like the holothosene (information-energy matrix left by previous patients) of that particular machine?

It is worth mentioning that the limitation of time has been always an important constraint (for all the 3 different series) as the time available was not enough to run as many sessions of the same type as it would be ideal, or changing subjects for all of them.

Experiments are still somewhat exploratory. There is so far no theoretical framework with some specific predictions to be tested. This certainly does not invalidate the experiments and their relevance. Many important scientific discoveries and progresses happened this way, like with the famous Maxwell equations, describing electricity and magnetism in a unified way, which could only be derived following a series of exploratory experiments done by Volta, Galvani, Faraday, Marconi, Hertz and many others. Quantum mechanics is another example of a theory that was created to deal with otherwise unexplained experiments, like the black-body radiation, the photoelectric effect and the electromagnetic radiation absorbed and emitted by atoms. And of course, physics is always facing new explanatory gaps, requiring new theories, as is the case for instance today with the unexpected experimental observation of the so-called dark matter and dark energy.

8 Future experimental steps

Based on the results obtained so far and also on the experience accumulated, the author plans to conduct the following steps, in short, mid and long terms:

o   Comparison with EEG experiments, under similar conditions (as described earlier, a pilot study is already on the way).
o   During the next fMRI experiments, to focus more immediately on effects of bioenergy on materials, not the brain, for initial simplicity and better control of the experimental parameters, to gather knowledge for better understand, later, about what happens in the brain, discriminating what is cause and what is effect.
o   Perfect protocol and technical conditions of the experiments, incorporating suggestions and criticism collected after publication of these preliminary findings and analyses.
o   Test different materials to find very responsive to bioenergy. Based on the egg experiment, it could start with pure proteins like albumin, collagen or laminin.
o   Use a 9T fMRI system (9 times stronger than the machine used in series 1 and 2, and 3 times the one used in series 3), to evaluate the effects of a more intense magnetic field over all the previous preliminary findings. For example: would it be even harder for the subject-energizer to move energies inside of such a machine?
o   Develop a protocol to measure intensity of the bioenergy based on the BOLD technique: standardize conditions, identification of an appropriate scale, quantitative analysis of the fMRI image, etc.
o   Test effects of: time decay of the bioenergy activation; bioenergy accumulation; distance between energizer and medium; different materials put on the way, between the energizer and

the medium.

o Expand the fMRI and similar experiments to a wider and more diverse group of subjects or practitioners, to find common characteristics and determine average values.

o Deepening of the research of other energetic maneuvers.

o Third-person experiments and research of out-of-body experiences.

o Broadening of the research via the study of the possible spontaneous occurrence of the VS in animals and the effects of bioenergy transmission to them.

o Comparative analyses (cross-analysis) of the fMRI study with the results of other similar researches (with the same objective, however, with different methodology), such as for instance, surveys, individual interviews, and bioenergy evaluation of the subjects by sensitive and experienced bioenergy workers (as the research conducted during IAC's course 'Goal: Intrusionlessness') (Trivellato & Alegretti, 2005).

o Experiments with the protein transducer; Reich based instruments; gas discharge devices (Korotkov devices), for comparison.

o Neurological analyses of the VS and bioenergy procedures through the use of other technologies, like, for instance:

  o PET scan (Positron Emission Tomography)
  o fNIRS (functional Near Infrared Spectroscopy)

o Analyses of other effects of the VS and bioenergy (endogenous or exogenous) on the soma:

  o biochemical changes: hormonal and metabolic
  o influence over the immune response
  o epigenetic changes, or the pattern of expression of genes (which ones would be activated; which ones would be deactivated; mechanisms of these changes)

o Development of a portable and simpler NMR device specific for bioenergy detection, measurement and analysis, as the size, price and complexity of the medical MRI system is a direct result of the need of a big hole in the main magnet for a person to be put in and also for the generation of 3D image for clinical diagnostics.

o Development of a Bioenergy Technology, analogous to the development of the mechanical, thermodynamic, electrical-

magnetic, and information technologies.
o   Test the hypothesis of ectoplasm as being an exotic state of the matter, as it seems to be many times created by bioenergy action.

9 Possible future applications

From the accumulation of data, the expansion of collection of case studies, the widening of the knowledge about the phenomena, and especially the establishment of average values and behaviors of the VS and bioenergy procedures through the examination of the greatest number possible of participants, it will probably be possible to develop the following practical applications, among several not yet glimpsed:

VS confirmation. External detection and confirmation of the VS in any person, including those that are still developing their specific parapsychomotricity, and therefore are not yet confident or lucid about their experiences, diminishing in this way the possible doubts about the existence or actuality of their VSs.

Energometry. Measure, indirectly through the neurophysiological measurement or directly through bioenergy transduction, of the power and broadness of the VS or the intensity of the transmission or absorption of bioenergy, also allowing the practitioners an initial feedback to facilitate their development.

Qualification. Analysis of the quality of the VS, through the indirect measurement of the attributes associated with its generation, such as: quantity of bioenergy, velocity, rhythm, broadness, cohesion, activation and others (Trivellato, 2008).

Intraphysicality. Determination of the degree in that a specific VS is physical (or has a somatic repercussion). It is anticipated here, for example, possible cases in which the participant has a VS that operates or happens mostly in other more subtle vehicles. In such cases, the cerebral analyses could indicate weak signals in where the participant could be convinced of having experienced an intense VS, but yet, subtle. The occurrence of a true VS could be confirmed by an external agent (researcher sensitive to bioenergies, able to measure the VS and its intensity), as to confirm the existence, in this case, of a VS with less interface or action over the physical vehicle, or through a direct bioenergy transducer, when and if available.

Mechanisms. Better understanding of the mechanisms and correlation of the factors that intervene in the VS and other bioenergy procedures: positive or negative; endogenous or exogenous.

Classification. Possibility of characterization and contextualization of different types of VS and bioenergy regimens.

Projectability. Possibility of detection of the imminent projection, when it is associated with the occurrence of the VS (a common condition for many lucid projectors). In certain cases, this detection could be used to generate the extraphysical awakening of the consciousness and to help in the obtaining of lucidity and control of the projection (in case the VS happens during takeoff), or yet to help stimulate the recollection of the projective experience (when the VS happens during the return to the soma). In other cases it could allow for the objective and technical study of the lucid projectability by researchers of Projectiology. A sensitive bioenergy transducer could in principle detect directly the presence of a person out of the body, for instance in a separate room from the latter, or by the bioenergies changes around his physical body. This could be the so much sought objective proof of the OBEs.

Biofeedback. The development of a supporting technology, in the form of biofeedback, which would facilitate the beginners to develop the ability to generate the VS or other bioenergy actions. In a way, compared to the topic 1, is like the relation that exists between medical EMGs and more specific instruments that help people to relax, or EEGs and the simpler instruments that can help people to enter and stay in alpha.

Training. Perfecting of the teaching method of the VS and other bioenergy techniques. All the findings obtained through this line of research would be used to improve the techniques per se and also the clarifications for their application, right from the beginning of the teaching process to the neophytes. Clearly it would help also those already working with VELO to perfect their personal technique.

Therapy. Improvement of certain therapeutic and self-therapeutic techniques. Given the importance of the VS and control of energies as holosomatic homeostatic resource (physical, energetic, emotional and mental balances), and as an energetic self-defense technique, it would be evident the application of the findings in this line of research aiming the integral health of the consciousness.

Bioenergy technology devices. Support to the development of a *Bioenergy Technology* with many possible applications, most still to be envisioned. It would contribute for research and development of bioenergy apparatus capable of interacting or dealing directly with bioenergies. According to the type of transduction, these applications could be divided in 3 groups, listed below, along with a few examples (here only considering, for simplicity, electromagnetism as the physical form of energy, as it is undoubtedly the most widespread in technological applications):

No transduction: absorbers, accumulators, conductors, insulators, switches, regulators, amplifiers

Transduction of bioenergy into electromagnetic energy: sensors or transducers capable of detecting and measuring bioenergy could be the basis of instruments such as: bioenergy-meters to assess peoples', animals' and plants' level of vitality and health; devices for medical analysis (equivalent to the Star Trek tricorder); bioenergy-meters for natural and artificial environments; more advanced means of communication, perhaps able to overcome the limitations of current systems; bioenergy imaging devices (cameras based on a "bioenergy CCD"); imaging of the interior of the body and even chakras, aura and meridians; remote sensing satellites for bioenergy geographical surveys; telescopes capable of taking pictures of the still invisible realities of the cosmos, perhaps even other dimensions; and instruments that could prove objectively the reality of out of body experience (astral projection), coming closer to the evidence of consciousness as other property of the universe, independent of matter.

Transduction of electromagnetic energy into bioenergy: devices capable of: generating bioenergy fields that can vitalize people, heal the sick, clean houses and environments; approximate dimensions or even build bridges between them; convert matter into ectoplasm, and vice-versa, enabling dematerialization, rematerialization, recycling of waste and other materials and even teleportation.

Paratechnology. The development of all these 3 forms of bioenergy technology could progress to a point in which it could work as an intraphysical interface with the extraphysical paratechnology.

10 Conclusion

The preliminary results so far are encouraging and point to real results as for the objective reality of bioenergies and their role in the consciousness' action over the brain and other forms of matter, and should be seen as meriting further experiments by other researchers looking for sustaining or falsifying these findings and their interpretation. It is very important to continue and progress with further experiments.

It is essential to emphasize that these results cannot yet be taken as definitive proof (if such a thing exists) of the objectivity of bioenergy and its manifestations. It is necessary that several other open-minded researchers replicate (repeat) these experiments. If they reach the same or similar results, then a greater acceptance of these findings would be achieved, first within the informed scientific community and then the general population.

If these findings are replicated and proven, they will cause enormous impact on physics, science in general and philosophy - that is, it would be evidence that consciousness is objective and independent of matter and is able to affect it and change some of its characteristics (even possibly at the level of protons and neutrons). This would inaugurate a new field in science and technology.

As discussed, the deepening of the study of the possible mechanisms of interaction between bioenergy and the BOLD and MRI techniques can inspire the development of better and more specific or dedicated systems for bioenergy detection and also for the creation of an explicative and predictive theoretical framework for the conceptualization and understanding of bioenergy.

Nonetheless, the main consequences would not be in the areas of technology, convenience or human comfort, but in the areas of philosophy, holistic comprehension of the deep nature of reality and of our understanding of ourselves as multidimensional consciousnesses. Perhaps it would even reinforce other evidences that humans (and other living beings) are transcendent consciousness, so that, similar to the NDE (near-death experience), it can help people to improve their worldview, consciential maturity, personal principles and collective ethics, ushering in a true golden age of human civilization.

Bibliography

Aerts, D. & Sassoli de Bianchi, M. (2015). Do spins have directions? Soft Computing 21, pp. 1483-1504.

Alegretti, Wagner (2008). An Approach to the Research of the Vibrational State through the Study of Brain Activity, Proceedings of the 2nd International Symposium on Conscientiological Research, Journal of Consciousness, Vol. 11, N. 42, International Academy of Consciousness (IAC), Portugal, p. 221. (Also published in this volume).

Alegretti, Wagner (1990). Tecnologia Bioenergética (Bioenergy Technology), Proceedings of the 1st International Congress of Projectiology, IIPC, Rio de Janeiro, Brazil; p. 32.

Ponhero, Rute Maria Rodrigues (2013). *Correlato eletroencefalográfico do estado vibracional* (dissertation for Physiological Psychology and Behavioral Studies master degree) – Universidade Federal do Rio Grande do Norte, Natal, Brazil.

Trivellato, Nanci (2008). Measurable Attributes of the Vibrational State Technique, Proceedings of the 2nd International Symposium on Conscientiological Research, Journal of Consciousness, Vol. 11, N. 42, Portugal; p. 168. (Also published in this volume).

Trivellato, Nanci (2015). Estado Vibracional: pesquisas, técnicas e aplicações (Vibrational State: research, techniques and applications), IAC – International Academy of Consciousness, Portugal.

Trivellato Nanci & Alegretti, Wagner (2005), Bases para o Energograma e Despertograma, Conference presented during the *I Jornada da Despertologia*; CE-AEC, 15-17 July 2005, Foz do Iguaçu, Brazil.

Vieira, Waldo (2002). *Projectiology: a Panorama of Experiences of the Consciousness outside the Human Body*, IIPC, Rio de Janeiro, Brazil, p. 22.

Note: This article was previously published in: Journal of Consciousness 18, No. 59, 2015, p. 517-549. An Italian translation is available in: *Studi sulla Coscienza*, AutoRicerca, Numero 10, Anno 2015, pp. 17-52.

AutoRicerca

Vibrational state: qualitative and quantitative analysis

Nanci Trivellato

Issue 20
Year 2020
Pages 155-190

Abstract

The vibrational state (VS) is a common spontaneous phenomenon correlated with out-of-body experiences. Endeavors have been made for the past four decades to devise techniques to produce the VS at will; however, practitioners' attempts to produce a VS seem to predispose misconceptions, especially among those who have never experienced a spontaneous VS. Aspiring to contribute to clarify this phenomenon, this paper presents some of the findings and conclusions from over 13 years of research on this topic, carried out using different methodologies — from self-research to evaluation of the effects produced in the energy field of 988 individuals and survey research with 676 participants from 31 countries. Careful analysis of collected data and reports from practitioners and experimenters was instrumental in the production of the *Scale of Vibrostasis*, which establishes the various degrees of activation on the energy body until reaching the intensity that triggers the VS, and the *Scale of Impact of the VS*, which helps assess the quality and intensity of the VS experienced. Both scales are presented here.

AutoRicerca, Issue 20, Year 2020, Pages 155-190

1 Introduction

Despite at least a hundred years' worth of reports, little is known about the vibrational state (VS), a phenomenon that has been modeled as an intense activation of one's energy body or biofield, producing an increase of its regular "frequency" of vibration that results in a resonance of all of its energies and energy centers (also known as chakras). Accounts from experiencers have often described it as a type of sensation that resembles each cell of their bodies vibrating powerfully, which they feel are produced by some sort of non-ordinary vitality or energy.

The VS is technically defined here as the specific energy regime characterized by the resonance, intensification, coherence, alignment and phase synchronization of all or most of the energy of the *energetic interface* (energy body) and, as a hypothesis, in some cases, the entire set of bodies (a hypothesized set of physical, energy, non-physical and mental vehicles of manifestation). It consists in a climax of a self-sustained and distinct vibratory regime of the energy that happens in unison. The VS can happen spontaneously or be induced at will. Among the characteristics and attributes of the VS are: intensity, scope, duration, frequency, completeness, and sustainability.

The technical name employed here for the activation (A) of the energetic interface and, consequently, for the level of activation that results in the VS, is *vibrostasis*.

Those who have experienced conscious out-of-body experiences (OBE, also referred to as lucid projection, astral projection, or astral travel) often describe they felt an intense vibration when they sensed they were projecting away from their physical bodies or upon perceiving a re-alignment with it. In some cases, this vibration is felt as or accompanied by a concomitant auditory perception such as buzzing or rumbling. Data from an online survey where participants were asked about sensations experienced in association with the OBE suggest that 56% of projectors experience vibrations (Buhlman, 2014) or intense energy-like sensations.

Note that, although the great majority of VS experiences occur in connection with the OBE, they can also happen in a number of different circumstances, including, in specific cases, during meditation, in acupuncture sessions, in deep conscious relaxation states, and similar conditions.

Explanatory Model

The model used here to explain the occurrence of the VS is based on the "activity" of the energy connection that exists between the vehicles of manifestation of the individual (self, consciousness).

OBEs can be interpreted by experimenters as a means to observe that we manifest via vehicles or bodies pertaining to different dimensions or different levels of reality. In other words, when experiencing a conscious OBE one will identify that a vehicle subtler than the physical body is being used for manifestation in the nonphysical dimension. Such paraphysical body (*nonphysical body*, or "astral body") appears to the senses of the experiencer to be connected to the human body (physical body) by a field of energy often referred to as the energy body (energetic interface or etheric double). The continuous conscious projection — i.e., being fully conscious and lucid during the three basic phases of the OBE: (1) while leaving the body, (2) during the nonphysical experience per se, and (3) when reentering the physical body — allows the observation of the energy connection between the two aforementioned bodies (nonphysical and physical body).

When in full alignment of the bodies, that is, in the intraphysical or physical state, such energy seems to cohere the physical and nonphysical bodies, permeating both vehicles and surrounding them, creating a type of field that is partially fused to them, and that binds them to each other. This theoretical field of energy is popularly known as aura and referred to as the *biofield* by the National Institutes of Health in the USA.

The fact that the VS often occurs in association with the projection of the consciousness, via one's nonphysical body, from the physical body suggests that the "stretch" of the energies of the energetic interface might trigger it, or vice-versa. According to the model adopted in this article, it is the activity or activation effect produced by the change in state of such energy connection that generates the VS, hence why most VSs are experienced during the phase of separation or reentering of the nonphysical body.

The occurrence of the VS while the person is in the regular waking state (completely aligned with the physical body) is also possible, although significantly less common than in the phases of disconnecting or reentering of the nonphysical body. Likewise, in some cases the VS is also experienced while projected, or "outside" the body.

In specific conditions, even in full alignment of the bodies, the energetic interface is activated or experiences an increased level of "vibration", which may result in the VS. Nonetheless, evidently, one may experience levels of increment of activation of the energetic interface that can be of lesser intensity than that required to generate a VS.

The looser or more fluid the energetic interface's energy is, the easier may be for the VS to be felt and for the separation from the physical vehicle (OBE) to happen. Rigid energies or areas with stagnated energies in one's energetic interface could result in reduced possibility of the occurrence of the VS, even during conditions that could potentially trigger it. These stagnant energies can be referred to as "blocks."

In the picture below, a visual representation of energy links that connect the physical and nonphysical body can be seen. As it seems, each of these links corresponds to a chakra, from the major to the minor ones (Figure 1), which integrate and promote the exchange of energy between these vehicles. Note: the illustration below is indicative only and does not represent all existing energy links.

The VS is experienced in the whole body, or rather, in the whole energetic interface, but is felt also in the physical body. However, the same intensity and type of activation or vibration can happen in only part of the energetic interface. In this case, it indicates that the resonance did not reach all the energies and, therefore, the parts not involved in this activation were (and remained) more stagnant.

Techniques to produce the VS at will, such as the classical voluntary energetic longitudinal oscillation (VELO) (Trivellato, 2008) are based on affecting the energetic interface so as to loosen its energies and eliminate blockages in its chakras or interchakral circuits (nadis), allowing for increased activity.

The set of all chakras forms the *chakral system* of the energetic interface, while channels of communication between chakras form the *nadic system*. The illustration in Figure 2 represents the chakras

(Trivellato, 2015) that are examined in more detail as part of this study. The names of the chakras are based on the physiology and anatomy of the human body in order to make them more universal and sensible.

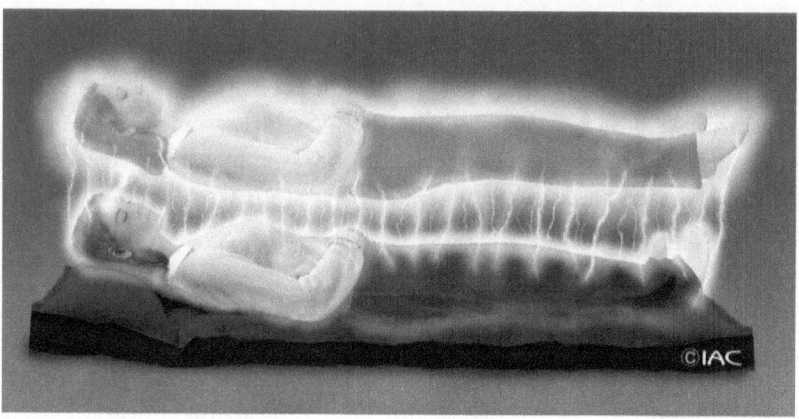

Figure 1 Representation of the "stretch" of the energetic interface, characteristic of the condition of small non-alignment of physical and nonphysical body or of the initial stage of the OBE, in which the energy links between both bodies are more distinctly perceptible. © International Academy of Consciousness.

As knowledge about the anatomy and composition of the physical body is in constant progress, similarly the anatomy and composition of the energetic interface (*energoanatomy*) needs to be investigated and developed continuously. As there is no way to "dissect" the energetic interface, nor does the appropriate technology to examine it exist to date, the identification of its constituent elements is based on theories and propositions derived from personal observations and experiences, some recorded since ancient times. This limitation or complexity in examining the energetic interface greatly hinders research.

Inherent complexities

Until recently, studies of the VS were based solely on reports of personal (subjective) experiences. However, proper understanding of a phenomenon, be it natural, physical, consciential, energetic, or psychic, requires the identification and, ideally, the measurement of said phenomenon ("operationalization" of the concept).

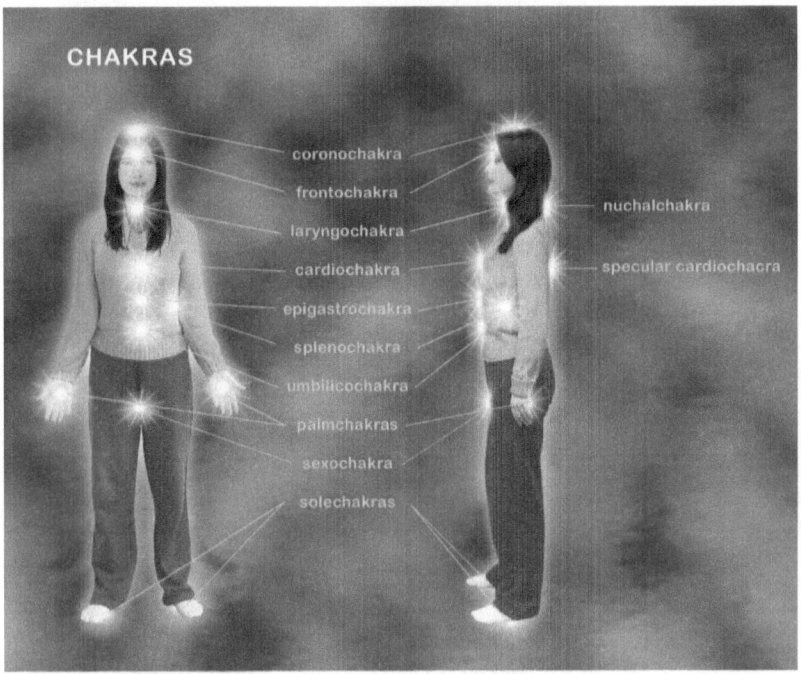

Figure 2 Representation of the main chakras and their position (Trivellato, 2015). © International Academy of Consciousness.

Some authors who studied OBE sensations would simply refer to the VS as "vibrations", "tremor", "frequency change" of the energy body, and so forth. Thus, many references to the VS have been made throughout history; but each individual has used different discourse and expressions to designate it and to describe its phenomenology. Without a name, a phenomenon is not properly acknowledged and, therefore, not properly examined. Having a shared language — specific language and definition — also makes it possible to establish discussions on the topic which, in turn, allows the information to develop through proper procedures and channels. Today, when analyses of OBE accounts are made, it is possible to conclude when those who experienced the OBE are referring to the VS because we know its characteristics and are able, therefore, to recognize its occurrence in reports from the past.

Once the name "vibrational state" was coined (Monroe, 1971) and accepted by a significant portion of the OBE community, accounts from experimenters that refer to it by name left little doubt about what experience or phenomenon was being described.

In regard to efforts made to evaluate the VS, researchers of the International Academy of Consciousness have been dedicating resources to find means to correlate measurements of the VS as well as other energy phenomena. Two major lines of research are well established within IAC: (1) second-person, inter-subjective measurement of the VS and of the attributes of the technique to produce it (i.e., VELO) performed by an individual via scanning and sensing the practitioner's energy during the experiment (Trivellato, 2008) and (2) third-person, objective measurement of correlates of the VS and VELO via examination of changes in the brain activity via instrumentation like fMRI (Alegretti, 2008).

Other efforts are also being made to find objective means to measure parameters correlated to the VS through instruments (technology and paratechnology). However, while such avenues are under development, we still must rely on a researcher capable to sense and measure the VS, be it one's own VS and energy activation (first-person research) or someone else's (second-person research), as in the case of the first line of research aforementioned. It is important to state that this type of research is not ideal, though, as it involves the researcher's subjective evaluation. With that said, having data generated from serious second-person research is far better than having none. What is critical in this case is that researchers who pursue second-person perspective research are properly trained and follow a pre-established measuring scale and evaluation criteria. It is important to remember that experiences themselves cannot be directly measured by devices and reduced to objective parameters, but the first, second and third-person approaches may be pursued in a complementary fashion.

Studying any type of subjective experience or personal feeling/sensation is in itself a very complex field for a number of reasons vastly discussed in social sciences. But such complexity is even more intense when it comes to the study of phenomena involving non-ordinary energy (*bioenergy*, subtle energy, vital energy, chi, or orgone), the energetic interface, and the non-ordinary or paraphysical reality (aka nonphysical reality). While identification of such complexity is needed, one cannot turn one's back to the fact that such experiences are lived and reported, and, therefore, must be acknowledged. Scientific investigations and their research methods

AutoRicerca, Issue 20, Year 2020, Pages 155-190

must properly take this complexity into consideration.

According to reports, the experience of the VS can leave a quite beneficial residual effect, producing some sort of internal balance or greater self-control. In addition, a number of studies about the impact of one's bioenergy on health and emotional state suggest that a healthier energy field can positively affect the individual (Trivellato, 2015). Such studies, together with the registered evidence and anecdotal observations, encourage efforts to devise effective methods to allow individuals to improve their bioenergy field in order to produce occurrences such as the VS; hence, the study on VELO.

It is important to stress that, despite the complexity in researching the VS and producing it intentionally, the sensation is unique and cannot be easily confused with other types of vibratory sensations or other bioenergetic phenomena. The intensity felt during the vibrational state and the ostensible sensations experienced can be so powerful, in some cases, as to generate surprise or fear to uninformed or inexperienced individual.

For those with no direct, personal experience of the VS, the best way to understand the sensations and impact of the VS is through examination of how some have described their experience. Below are a few examples, listed in alphabetical order of author.

"In seconds I'm shaken awake by intense vibrations and a roaring sound throughout my body. It feels like I'm in the middle of a jet engine and my body and mind are about to vibrate apart. I'm shocked and scared by the intensity of the vibrations and sounds and snap back into my body." *William Buhlman* (1996, p. 8).

"I heard also, when in this state, in addition to physical sounds, several strange noises: crackling sounds suggesting electrical phenomena; roaring and whirring noises as of gigantic machines; a peculiar snapping sound, recalling the driving bands, used to transmit power in a workshop; sounds like the surging of an angry sea and rushing winds; and sometimes voices calling. Some of these sounds may have been caused by variations in blood pressure, but I do not think that all of them can be accounted for in this way." *Oliver Fox* (1962, p. 62).

"It was not a shaking, but more of a "vibration," steady and unvarying in frequency. It felt much like an electric shock running through the entire body without the pain involved. Also, the frequency seemed somewhat below the sixty cycle pulsation, perhaps half that rate. Frightened, I stayed with it, trying to remain calm. I could still see the room around me, but could hear little above the roaring sound caused by the vibrations." *Robert Monroe* (1971, p. 24).

"[…] I thought it was my physical [body], but it was my astral commenced vibrating at a great rate of speed, in an up and down direction, and I could feel a tremendous pressure being exerted in the back of my head, in the medulla oblongata region. This pressure was very impressive and, came in regular spurts, the force of which seemed to pulsate my whole body." *Sylvan Muldoon* (1929, p. 51).

"A kind of electrical 'vibration' violently swept into my body, filling it with an electric-like shock and a terrible roaring noise. I thought I was being electrocuted and my first reaction was sheer panic." *Robert Peterson* (1997, p. 16).

"All of a sudden my body started to vibrate with an unbelievable energy, and successively a deep pain and sadness escaped from me. The energy circulating in my body, and throughout my hands was so intense that I could not stand, and I fell on my knees. I was grateful for this gift of energy, which allowed me to free myself from my pain and sadness. […] An amazing vibrational form of energy was crossing my entire body. I didn't know how it looked from the outside, but my impression was that I was about to dematerialize, and asked myself if the other persons in the room were still able to see my body. […] The phenomenon lasted about a quarter of an hour, with the vibrations gently diminishing, leaving me a bit dazed and without any hunger that evening." *Massimiliano Sassoli de Bianchi* in: (Trivellato, 2015; foreword, p. 1).

"Immediately there came over me a powerful tremor, from the head and over the whole body, together with a resounding noise, and this occurred a number of times. I found that something holy had encompassed me. I then fell asleep, but about twelve, one or two o'clock in the night there came over me a very powerful tremor from head to the feet, accompanied with a booming sound as if many winds had clashed one against another. It was indescribable, and it shook me and prostrated me on my face." *Emanuel Swedenborg* in: (Van Dusen, 1981, p. 43).

"The first VS experience I can remember happened when I was about

10 years old. I woke up in the middle of the night with the impression that the house was shaking, as in an earthquake. I felt a distinct and intense sensation, as strong vibrations and currents took over each minimal segment of my body." *Nanci Trivellato* (2015, p. 63).

"In one of the strongest VSs I had, I experienced complex — and very hard to describe — sensations, which were accompanied by unparalleled awareness. The sensations encompassed the perception of an internal vibration and a feeling as if each molecule of my body were alive and 'bubbling'. Such vibrations were not neutral, as they were like a pleasant 'agitation', full of vitality and vivacity. They came simultaneously with the clear recognition of the physical, energetic and nonphysical bodies separately, and a sense of cellular and para-cellular invigoration. All that seemed to amplify clarity and aware-ness, resulting in a state of ecstasy and striking wellbeing that lin-gered for days after the event." *Nanci Trivellato* (2015, p. 64).

It can be observed from the above accounts that, even though the nature of the energetic activation in the VS is very specific, different degrees of intensity and, consequently, different types and duration of effects can be experienced from each VS.

Probably the most challenging part of the research reported here was to evaluate the individuals' experiences — both the narrated ones and those examined through second-person measurements — and make sense of the expansive and complex data, yielding a scale that would reveal the progression of the VS. The scale of intensity of the VS presented here is called Scale of Impact of the Vibrational State, as it not only measures characteristics of each VS experi-enced, but also perceived effects.

To offer a general description of what the VS feels like, in my recently published book, I wrote:

"In an attempt to perform the complex — and perhaps unfeasible — task of providing a description of the feeling of the vibrational state, I would say it feels as a powerful vibration throughout the body, as if all cells were in a frenzy of pleasant and vivacious intense activation, bringing awareness. Such vibration can even be inter-preted by the less experienced individuals as physical vibration. It is often felt as if the body were connected to a high voltage electric current, so palpable that, in some cases, it could be 'heard.' The sensations arise, in general, from the perception of coherent, phased vibrations, sometimes described as widespread internal chills that may produce cold or heat sensation. Descriptions may also be made by use of analogies such as overloaded transformer; turbine engine; nuclear reactor in full swing; glaring incandescence radiation; pleasant fire; tuning fork; internal dynamo; live pulsating waves; *zillions* of pleasant needles in the body; and resonance" (Trivellato, 2015, p. 69).

2 Research methods

The information subsequently presented is the result of different investigative projects, based on the following methods, which include first, second and third-person perspective research:

- **Hetero-energometry** – one of the major components of this investigation, the hetero-energometry, is based on the "measuring" of the energy condition of other individuals who were submitted to evaluation when performing the technique to install the vibrational state (VELO) and had their energy field examined in the moment they claimed to be experiencing the VS. It is worth referring to the above discussion, which acknowledges that this research methodology (second-person perspective) is complex in itself due to the fact that the evaluating agent (the researcher) must use his/her own sensitivity — hence, subjective — to measure said energy effects. Several "calibrating" efforts were made, including comparing measuring results and impressions with other (equally trained) researcher. The outcome from these efforts strongly suggested that data resulting from this line of research have merit.

- **VS Survey** – a vibrational state survey carried online.

- **Literature review** – examination of relevant books related to the VS and its association to the out-of-body experience.

- **Case study** – specific experiences that had been found in the literature or that had been reported to me directly were examined in search for more detailed and comprehensive information.

- **Analysis of accounts and questions occurring during events** – the vast anecdotal collection of comments from participants in lectures and courses, either specifically about the VS or related to bioenergy in general, that I gave for the past 25 years was taken into consideration when performing the study.

- **Self-research** – self-energometry or, in other words, my own experiences and how I sensed and measured my VSs as well as lower levels of energetic activation.

3 Discussion

Details on two of the major research components that yielded data for this paper are presented below.

Hetero-energometry (individual energy evaluations)

One-on-one sessions aiming at measuring individuals' bioenergy effects by means of hetero-energometry represent a key component of this study. Some of the groundbreaking knowledge derived from these sessions is published in the *Journal of Consciousness* volume 11, number 42 (Trivellato, 2008).[1]

The continued collection of data and detailed examination of results from the individual evaluations carried out since 2008 produced the information presented here, as they favored the understanding of the progression of energy activation to produce the VS, the misconceptions about it, and the bottleneck for producing it. Such results are based on data collected for over 13 years, having 988 subjects examined through 2,342 one-hour sessions. Evaluation sessions took place in seven countries (United States, Portugal, England, Japan, Netherlands, Spain and Brazil) and included participants from many different cultures and backgrounds. At least 518 of the subjects were evaluated during three to four sessions carried out throughout approximately a one-year period.

---

[1] The journal can be accessed online at: *www.JofC.org.*

In a number of occasions throughout the data collection period, the main researchers who carried out the individual evaluation sessions, Trivellato and Alegretti, compared their results to cross-reference their observations on the same subjects to ratify their reliability and to calibrate the measurement procedure. Consistent perceptions, matching interpretations, and similar observations between them contributed to the motivation to carry on the study.

The surprising information gathered from those evaluations revealed that fewer individuals than expected were able to produce the VS at will, while some who were claiming to install the VS at will did not demonstrate understanding of the phenomenon. Most of the participants were incapable of getting close to the level of energetic activation necessary for it to be considered a VS.

Closer examination of the results revealed that 95.7% of the subjects examined did not reach a VS of any intensity during the evaluation session(s); however, an even more serious observation was that the great majority of them considered an estimated 20% to 30% of activation level of their energies to be equivalent to a VS. One point in common among those individuals was the fact that they learned how to produce the VS through the literature by Waldo Vieira (1986, 2002) or were influenced by his teachings throughout the last two decades or so. Vieira emphasized it was critical to achieve the VS at will, any time and multiple times, for someone to be considered balanced and self-determined. As a result, individuals felt they *had to* achieve it. Such affirmations that only those who produced the VS at will whenever and wherever they wanted were in good condition caused pressure on individuals who felt obliged to be able to self-induce the VS — a goal that is not always easily achieved, as will be discussed in this paper. In order to accommodate this requirement to the community around him, in the last at least 15 years, the notion of VS began to be distorted in Vieira's speech so that it would be more feasible for individuals to generate the (supposed or pseudo) "VS". So, in a large group of those who had followed Vieira's conscientiology framework, the grasp of what a VS actually is and the complexity to produce it at will in the waking state were lost throughout the years and, gradually, more confusion than clarification about the phenomenon was propagated (Trivellato, 2015).

Breaking such a "culture", which has grown among those who developed and learned about VS from this perspective, has proven a challenging undertaking. Studies such as this, carried out under IAC's auspices, aspire to foster better understanding about the phenomenon, which, in turn, can correct the misconceptions that now exist about the details of controlling and generating the VS at will.

Instructors who have been teaching the VS based on Vieira's view ended up, therefore, disseminating information that was partly useful, but also partly very confusing. At least this is the case of the universe of individuals who had been examined in the hetero-energometric research or responded to the VS Survey.

The groundwork investigation and findings that first revealed such a scenario were published in the paper *Measurable Attributes of the Vibrational State Technique* (Trivellato, 2008).[2] Since the identification that the pedagogical methods used to teach the VS could be weakening the possibilities of real understanding and controlling of the phenomenon (and, maybe, were even encouraging banalization of the real VS), the IAC has strived to establish well-founded information on the phenomenon and devise adequate pedagogical strategies to support those who want to research or experiment the VS. Some of those strategies are discussed in the said paper.

Devising a sensible technique that has higher chances of producing results and is explained in detail is key, as this could increase the number of legitimate VS experimenters (potential self-researchers), thus facilitating more and diverse formal research projects and, subsequently, scientific dialogue. However, it is not the objective of this paper to discuss techniques.[3]

Among the objectives of this work is to present a scale that shows the level of increase of the energetic activation until the vibrational state is installed. Up until recently the knowledge available was a simple description of the VS, which did not explain how it is installed and did not cover the intermediate levels of activation. In other words, activation was described as a quantum step or as a binary condition: one would either experience a VS or nothing noteworthy, as only those two points in the scale were contemplated. It is my view that this lack of recognition of the buildup steps to

---

[2] Re-published in this volume.

[3] The attributes of the VELO technique are discussed in the aforementioned paper. In addition, painstaking descriptions of each step and aspect involved in the execution of VELO, which seems to be an efficient method, is provided in the book *Vibrational State: research, techniques and applications* (Trivellato, 2015).

producing the VS is part of the reason for the distortions of the VS concept, as, upon applying the technique, one may experience positive effects even without achieving the VS. So, even if the purpose of the exercise is to produce the VS, in spite of the fact that the result may not be a VS, it can still be more than "no result" or "no sensations" whatsoever.

The Scale of Vibrostasis (vibroscale) presented in the next main section, which was developed based on the aforementioned research, aims to assist experimenters in evaluating or grading the intensity of what they experience. The existence of such a scale is a first step towards establishing a vocabulary with which the experiencer can acknowledge the intermediate effects of their energy practice and providing a shared language to allow and encourage discussions and further study.

The VS Survey

The survey was devised and developed in 2012. After a phase of testing and debugging, invitations to complete it was made to the general public in November 2013. Data was collected for 14 months and a total of 676 participants from 31 countries replied to the questionnaire.

Any individual who had ever heard of the vibrational state could answer this online survey. Call for participation was done via available means at IAC, including public Facebook posts both in IAC's page as well as in other organizations page, and YouTube videos. Information was also sent to IAC's database.

As the graph below shows (Graph 1), the overall result obtained from the survey, regarding how participants experienced the VS, reveals that 18.5% of respondents only experienced spontaneous VSs while 39.3% claimed they only experienced self-provoked VSs through techniques.

The VS Survey had two key objectives. The first one was to evaluate the phenomenology revolving the VS, including the most common sensations experimenters associate to it. The second one was to verify the experimenters' level of understanding of the VS via observing the consistency of their answers.

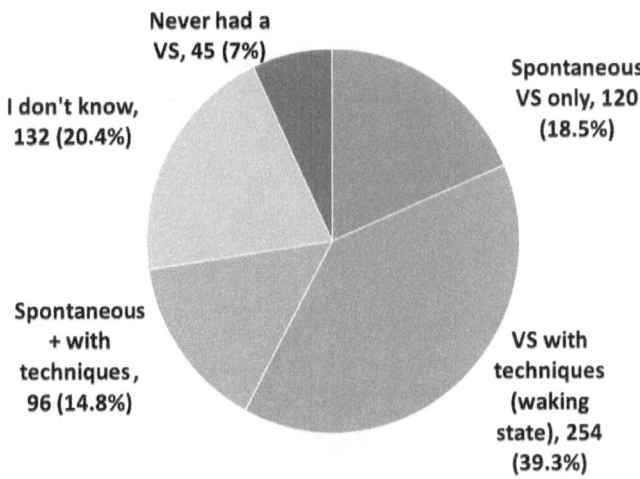

Graph 1 Experience of the VS.

A number of technical measures were performed in order to avoid inconsistency in the answers; nonetheless, some questions purposefully allowed them with the aim to measure how clearly individuals really knew what the VS is and how much they were consistent in describing their own sensations.

The data derived from the VS Survey revealed a significant percentage of ambiguous or contradictory information, which makes it difficult to confidently draw substantial conclusions from it. For instance, someone who declared never to have had a VS was not expected to answer questions in which they were asked about the sensations they felt when experiencing a VS. Likewise, someone who claimed to have control over the installation of the VS would not be expected to describe their VSs as always occurring spontaneously or to report sensations that are patently not related to the phenomenon. Speculating about the possible reason for such inconsistent answers, it is possibly due to the fact that individuals are packed with contradictory theoretical "teachings" about what the VS is and feel pressured to perform highly according to the requirements established by such teachings, as pointed out in the discussions in the section "hetero-energometry" above. Note that over 97% of survey respondents learned about the VS directly or indirectly through Vieira's approach.

As per the graph below (Graph 2), only 23.5% of the respondents pointed out sensations of their VS which were consistent with the nature of the energy activation that occurs in a VS, while approximately 40% referred to sensations that are contradictory to a VS occurrence:

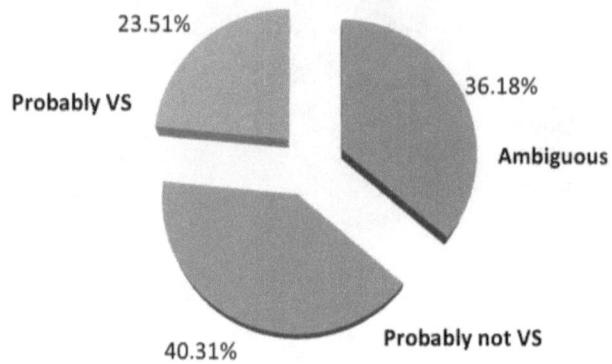

Graph 2 Coherence in the description of claimed VS sensations.

One point to take into consideration when examining such contradictory results is the fact that, until very recently, serious studies on the VS were virtually nonexistent. Data retrieved from printed and electronic literature on the subject up to the point of this study showed that findings about the VS were drawn from questions that usually appeared on surveys dedicated to examining the out-of-body experience. Hence, the prior research only examined reports and descriptions of the VS found, in most cases, in accounts of OBEs.

Hopefully, this research paper, together with larger publications dedicated exclusively to the VS, will bring more understanding about and awareness of the phenomenon and provide better means for triggering it at will and for examining it. A second wave of the VS Survey, with new structure and strategy, is planned for 2016 when, possibly, participants will be better informed about the phenomenon and with better self-energometric procedures.

The most coherent answers found from this first VS Survey seem to be from those who had experienced a spontaneous VS. Of the 499 individuals who declared having had a VS, 220 said they had spontaneous VSs, in association with the OBE. The most common description of the VS these individuals reported was: "intense sensation of electricity throughout the body, which remains spontaneously strong and unquestionable for a certain period of time."

A vast majority (72.1%) of those who declared to have had spontaneous VSs informed they occurred in a state of deep relaxation (Graph 3). This reinforces experiential evidences that achieving a true VS in the normal waking state is not as easy as one may think. This is also consistent with the observations made in the individual evaluation sessions aforementioned as well as with personal experience. Installing a VS at will requires practice, experience, and knowledge of the technique, processes, and variables involved.

It is important to mention that, in exceptional conditions, when the relevant energetic, internal/intraconsciential, and environmental variables are right, a VS can be achieved very easily, sometimes only by relaxing or desiring it. Even though in some of those cases there may be some level of will involved, this event cannot be considered a "technique" and is not the standard occurrence.

When those who had had VSs both spontaneously in relaxed conditions such as in connection to the OBE and in the waking state were asked how the sensation of the VS experienced in deeply relaxed condition compared to the sensation of the VS produced in the waking state with a technique, 80.2% of the respondents coherently replied that the nature of the sensation was the same, however, the intensity was lesser in the VSs generated at will (Graph 4).

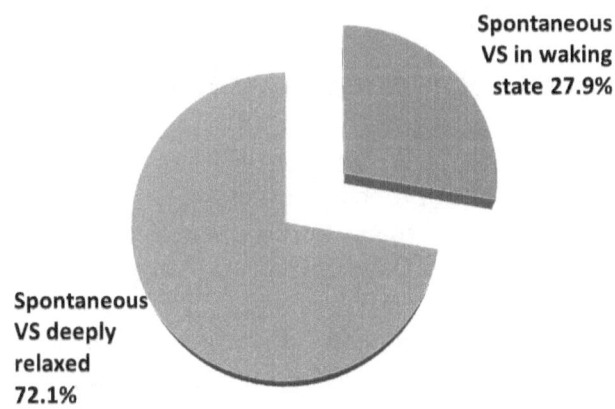

Graph 3 State when the VS occurred.

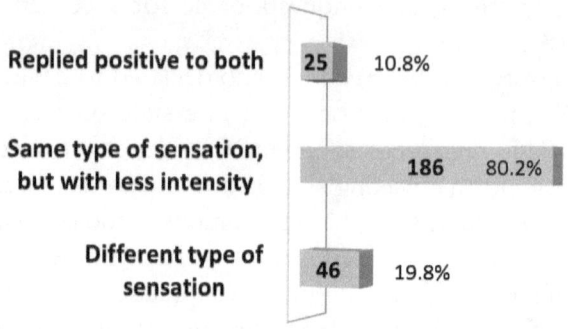

Graph 4 Comparison of the sensations of spontaneous VSs in association with OBEs and VSs produced at will in the waking state

This result points to the need of increasing the level of education of practitioners about the VS so that those who have not yet reached the VS will know they must continue investing in loosening and activating their bioenergy to be able to trigger the VS. Logically speaking, an occurrence that produces completely different effects, having different nature, different aspects and different sensations from the studied phenomenon, is most probably a different phenomenon. So, having a clear notion and characterization of the VS should help reducing confusion about it. The Scale of Vibrostasis below is an effort to contribute to such step.

4 Scale of vibrostasis (vibroscale)

Vibrostasis is the term proposed to designate the different forms, intensities and modes of oscillation or vibration of the energy body, in different levels of activation, which can progress until it culminates in the VS.

Considering that the exponential increase in activation leads to the energetic dynamism that produces the VS, vibrostasis is a key element to be understood to analyze the progression of this energy effect. Note that groundwork information on vibrostasis was published under "Activation (A)" in the article *Measurable Attributes of the Vibrational State Technique* (Trivellato, 2008), as *activation* was the general expression used until recently to refer to this concept.

The emergence of the VS phenomenon is characterized by and experienced as an intense vibrostasis. However, when the practitioner is trying to produce the VS at will based on techniques such as VELO, the increase in the activation of the energetic interface may not be enough to reach the minimum level of activation (A_{min}) necessary to trigger the VS. It is necessary, therefore, to have a scale of activation so that the practitioner can assess how close (or far) from the VS he/she is when performing the technique.

Identifying the level of activation reached when performing bioenergy techniques is a form of energometry. Communicating one's experience based on an established scale of vibrostasis should naturally foment discussions and further investigation on bioenergy effects, as experimenters and researchers will have a common terminology that is as unambiguous as possible.

It is worth mentioning that there are several energetic effects one may experience when performing bioenergy practices. Among them is the activation, from which other effects may also derive.

The scale

The scale of vibrostasis (or vibroscale) is the scale of activation of the energetic interface. Below are descriptions of five vibrostasiological classifications, proposed as a tool to assist assessment of the energy effects experienced by the experimenter or practitioner, according to the energy activation level or the vibrostasis experienced.

Level 0 (zero), of course, does not exist in the proposed scale because there cannot be an individual with zero energetic activation, as it would mean absence of life or, in the case of the consciousness, "inexistence". So, when a practitioner self-designates "zero" as their perceived result, it would denote the practitioner's inability to perceive his/her own bioenergies, to master them, or to identify the effects produced by them.

As is to be expected in this type of graduated numerical scale, self-assessment of the practitioner presents a degree of subjectivity, as it will depend on his/her sensitivity, points of reference, and experience. Another factor to take into consideration is that the basal level of activation (defined below as naturovibrostasis) is particular to each individual, according to their general energetic condition, including their energetic looseness.

Seeking to produce a vibrostasiological scale as reliable and universal as possible, great attention was placed to the research data that gave rise to such scale aiming at recognizing patterns and leveling them (weighing and normalization of occurrences and experiences).

With the development of this area of study (vibrostasiology), it is expected that the expansion of knowledge allows for the improvement of such a scale and, perhaps, even the inclusion of new intermediate grade points.

(1) Naturovibrostasis

- From Latin *naturalis*: natural, innate + vibrostasis;
- Corresponding to up to 20% of energetic activation;
- Natural, ordinary state of bioenergetic activity associated with organic life;
- Basal vibration of the energetic interface;
- Monovehicular vibrostasis (energetic interface only).

In most cases, individuals are not aware of their naturovibrostasis. Even if their level of sensitivity and multidimensional self-awareness were to allow such discrimination, permanent and stable sensations, in general, become "ignored" by the brain, as is the case when we fail to realize the bloodstream flow or the heartbeat under normal rhythm and intensity. Given their stability and constancy of these sensations, we do not realize them, but if there were a change in them, they would be noticed.

However, an individual with good proprioception (ability to sense one's own physical body) and focused attention may notice the existence of such sensations, even under normal conditions. Likewise, by focusing their attention on their energetic interface and exerting action over it (i.e. moving the energies), practitioners with some energetic sensitivity and energoproprioception (ability to sense one's own energy body) could identify the naturovibrostasis. But, in this case, the fact that the sensations of the naturovibrostasis were only identified after paying attention and carrying out an energetic action is due to the factors mentioned above (becoming aware of the natural condition that was already there), therefore, one should not confuse such sensations with an increased level of energetic activation.

Usually, the identification of the naturovibrostasis can occur during VELO, but is more commonly perceived upon stopping it, when attention can be fully dedicated to "feeling." During the VELO exercise, some practitioners still with little sensitivity or difficulty of concentration and perception probably will have less awareness of naturovibrostasis, because (1) it is still a weak activation and (2) the practitioner has his/her attention highly focused on the control of energy pulses through the energetic interface, therefore, the attention is not directed only at noticing the sensations. This explains why some practitioners interrupt their VELO to "feel" or check the energy effect obtained until that moment (an undesirable action, as this ceases the progression of the activation). However, with the development of the practitioner, it is possible to be focused on VELO and record, at the same time, the energetic effects and sensations.

A practitioner should value his/her ability to feel naturovibrostasis, since it is his/her way of perceiving or identifying the energetic interface, being, therefore, an important element in the recognition of his/her parapsychic style, but should be aware that this feeling is not a result of a VS occurrence.

Thus, in the process of achieving control over the activation of their energies, the stage of self-research is inevitable and practitioners should regard it as a privilege and an exciting activity. As a researcher delights in analyzing data and discovering yet more complexity in the information obtained from his/her experiments, or in identifying contradictions in the data, thus, being more knowledgeable of his/her object of study, the VELO practitioner should show motivation and disposition toward his/her self-research.

(2) Leptovibrostasis

- From Greek *leptós* (λεπτός): weak, subtle, thin, small + vibrostasis;
- Corresponding to an energetic activation of 20% to 40%;
- Subtle increment of activation, still superficial, with little change in the natural, basal energetic state;
- Predominantly monovehicular, but with some physical effects as well (sensations and effects start to reach the).

Normally, leptovibrostasis produces milder sensations and effects, not getting to the point of producing deeper and longer lasting changes in the overall condition of the experimenter.

According to the analysis of individuals tested for VS in hetero-energometric research, it was observed that the great majority of individuals considered this activation range (even at a low level, i.e. 20% – 30%), erroneously, to be the VS.

Such a misconception is natural in the case of someone who has never experienced a VS and is lacking, therefore, personal points of reference. Hence the importance of understanding what the VS is, as it reduces the occurrence of this misidentification.

Misinterpretation is also common in the case of someone who has spent years trying to reach a VS through techniques with no result beyond this level of vibrostasis, especially for those who felt they "necessarily had to produce VSs at will", as is the case of many of those examined in this study as previously mentioned. Over time, the individual "gives up" and concludes that "it was only that effect that was supposed to happen"; in other words, convinced oneself that the leptovibrostasis is the VS.

(3) Midivibrostasis

- From Latin *medius*: middle + vibrostasis;
- Corresponding to an energetic activation of 40% to 60%;
- Level of activation equidistant or intermediate between the minimum and maximum points in the scale, i.e., naturovibrostasis and holovibrostasis;
- Bivehicular vibrostasis (involves both the energetic interface and the physical).

This is the stage of activation in which energetic effects obtained by VELO are perceived more ostensibly. The following results, among others, begin to be perceived in a more evident way from this point of the scale onwards: (1) energy uncoupling, which is the breakage of undesirable connections with the energy field of others; (2) reaching of energetic layers[4] or areas that still have stagnant energies; (3)

---

[4] The expression "layer" is employed here in a more intuitive than literal meaning, since the energetic interface is not formed like an onion. The use of this expression is an endeavor to communicate the notion of position in space in terms of the energetic interface as well as of age and severity. Hence, defining an energy as "deep" or "superficial" also refers to its dimensional or vibratory level. So, one may have energy pockets where energies still reside in stagnation, forming a "nodule" that is not easily achievable. Energy pockets may even be composed of ancient energies, such as from experiences lived before the present life, also causing them not to be necessarily readily observable and reachable. The idea of energy being in a deeper layer may, therefore, bring the notion of something chronic, deep-rooted,

identification of blocked chakras and, therefore, of energetic impediments for the energy flow; (4) relative increase of one's level of free conscential energy (CE$_{FREE}$) (Trivellato, 2008).

A significant percentage (approximately 30%) of those who were properly trained in the VELO technique came to perfect their VELO to the point of producing midivibrostasis. This corresponds to approximately 85% of those who diligently dedicated to practicing during the period they were guided into their exercises and followed in the longitudinal heteroenergometric examination. In the case of this group of people, about 40% were able to perceive and recognize the sensations that are characteristic of this level of activation of the energetic interface.

(4) Hadrovibrostasis

- From Greek *hadrós (ἁδρός)*: strong, evident + vibrostasis;
- Corresponding to an energetic activation of 60% to 80%;
- Intense energetic activation which, when producing chakral resonance, triggers the VS;
- Predominantly bivehicular (energetic interface and physical), but with some participation of the nonphysical body as well;
- Corresponds to verovibrostasis (from Latin *vero*: real + vibrostasis) of the following strengths:
- VS of intensity level 1 (I$_{VS}$1): 60% to 70% activation;
- VS of intensity level 2 (I$_{VS}$2): 70% to 80% activation.

When hadrovibrostasis is reached, there is the intense, complete resonance of the energetic interface, at a peak, typical of the VS. When experiencing for the first time a level of vibrostasis near 80% or greater, some individuals who have never heard of the VS may wonder if there could be something "wrong" with their bodies, due to the powerful and unusual sensations.

(5) Holovibrostasis

- From Greek *hólos (ὅλος)*: whole, with all parts + vibrostasis;
- Corresponding to an energetic activation of 80% to 100%;

---

old, or well-hidden in the energetic interface's systems, being the opposite of new/renewed, loose, or not yet well established energy, which would be referred to as more "superficial".

- Intense, usually associated with the non-alignment of the bodies or partial disconnection of the nonphysical body, being more commonly related to the departure or return to the physical body in an OBE;
- Trivehicular vibrostasis (involving energetic interface, nonphysical body, and physical), allowing not only the identification of the energetic effects but also the physical and nonphysical ones as well;
- Corresponds to verovibrostasis (from Latin *vero*: real + vibrostasis) of the following intensities:
  - VS of intensity level 3 ($I_{VS}3$): 80% to 90% activation;
  - VS of intensity level 4 ($I_{VS}4$): 90% to 100% activation.

Vibrostasis typical of the occurrence in the condition of non-alignment of the bodies, which causes an intense and potent activation in the energy body, but whose effects are also noticeable beyond such body.

Level of intensity harder to attain through techniques in the physical waking state. As mentioned before, results from the VS Survey showed that approximately 80% of participants who said they had experienced the VS in non-alignment conditions claimed that the VSs produced by them through techniques in a state of alignment of the vehicles were less intense.

The holovibrostasis can change experimenters more profoundly, including their inner state and the energetic or nonphysical connections they may have, reducing intrusions and reception of undesirable energies, having multivehicular effects far more lasting (in some cases, even permanent) than those in the previous levels.

In the case of lucid takeoff of the nonphysical body (in a lucid OBE), experimenters can also clearly identify the effect of this energetic activation in such nonphysical body. In addition, they can clearly feel the effects of the VS in their physical body as well, evidently stronger than experienced in the less intense levels of activation.[5]

Those who have experienced this top of scale or apex vibrostasis take it as their "reference VS", or the VS to which all others are

---

[5] Further studies to determine the occurrence of changes in the physical, whether physiological or other, still need to be conducted. The possibility that the VS causes direct effects on the physical body has been object of study in research carried out by the IAC.

compared in order to grade the intensity of each vibrostasic result.

In the self-sustained holovibrostasis — as is usually the case, for example, at the time right before lucid separation of the nonphysical body — in some instances practitioners may be able to act so as to extend the length of the VS or change its pattern (frequency, modulation, intensity, or other), adjusting it for different purposes, even to expedite the separation. This type of control of the VS seems to be performed by some form of modulation of its amplitude or frequency.

In some cases, the intensity of the holovibrostasis can produce such an intervehicular effect that it favors the "looseness" (liberation) of the *mental body* (mental core), hence, favoring equilibrium, expansion of lucidity, and integral balance (VS of tetravehicular effects). Such a rare occurrence can eventually lead to cosmoconsciousness (e.g., enlightenment, samadhi, satori).

5 Scale of impact of the vibrational state

Depending on the intensity of the vibrational state (I_{VS}), different effects and outcomes are experienced. The scale of impact of the VS takes into consideration aspects and attributes of the VS taxonomy presented in the book *Vibrational State: research, techniques and applications* (Trivellato, 2015). The scale aims to take into account (1) the aspects that are relevant in characterizing the I_{VS} and (2) the type of reaction and result that could aid in establishing it.

Below one can appreciate the key aspects that have been identified, researched, and catalogued from the large data derived from this investigation, especially through hetero-energometry as above discussed and from self-energometry as well.

- *Alignment* = degree of non-alignment of the bodies
- *Energetic uncoupling* = quality of the disconnection from unwanted energy fields or consciousnesses previously attached or linked to the experimenter
- *Consciential condition* = general changes in one's mental state, energy health, personal feeling, and physical condition

- *Deintrusion* = type, duration, and quality or extent of the completeness of the deintrusion (removal of unwanted nonphysical presences) produced
- *Duration* = how long it lasts
- *Duration of activation* = for how long the activation effects can be experienced
- *Duration of results* = how long the benefits last
- *Effects on mental body* = how much the mental body participates in or is affected by the VS
- *Effects on paragenetics* = what are the chances of transformations to occur in the nonphysical body (nonphysical genetics)
- *Energetic detoxification* = level of purging of foreign, toxic energies
- *Energetic fluidity result* = how it affects one's energetic fluidity
- *General parapsychism* = type, quality and degree of increment in one's psychic perceptions and abilities
- *Homogeneity* = how it distributes throughout the bodies
- *Intensity of sensation* = type and clarity of sensation
- *Intrusionlessness (prophylaxis and defense result)* = how much the VS impedes the invasion (clear or disguised) of undesirable foreign energies, ideas or feelings
- *Paraproprioception* = level of increment of one's perception of his/her subtle energies and subtle bodies
- *Personal reaction* = typical reaction of the experimenter
- *Predisposition for energetic activation* = effect that may facilitate future occurrence(s) of high vibrostasis
- *Projectability* = what effect the VS has over the level of "looseness" of one's bodies and one's capacity to project
- *Reach of effect* = depth (energetic layer) that the activation reaches
- *Robustness (of the VS)* = Level of endurance against interferences; resilience of the VS
- *Self-defense* = level of self-defense obtained
- *Sustainability (of VS)* = capacity of the activation to hold itself (degree throughout time, even after the VS)
- *Sympathetic deassimilation* = level of nullification of superficial

and deep sympathetic assimilations (i.e., of sensations and effects that have been "captured" from others) produced
- *Unblocking* = level of energetic unblocking produced
- *Vehicle* = sensations in relation to the body(ies) that is(are) most affected

The scale of impact shown below is applicable to the analysis of the characteristics of each VS experienced, not the condition of the experimenter. Therefore, listed effects and characteristics reflect results that are possible with the occurrence of one VS in each respective level, except when otherwise stated.

The description provided for each level of VS in the table that follows may assist experimenters in examining each VS occurrence they have and determining its quality and impact.

When at least 70% of the characteristics listed under a certain level have been experienced, it means the VS is most likely of that level, unless a specific parameter of higher weight related to another level is experienced. For example, in the case of the occurrence of an intense and multivehicular VS which provokes the continuous conscious projection/OBE this characteristic supersedes other less notable ones.

The results of a certain level of VS can be present in a VS of a higher level, as there is a cumulative effect. Thus, a VS of intensity 4 (Ivs 4) can include the results corresponding to the levels 1, 2 or 3.

Evidently, this scale mixes different types of VSs, including those that are self-induced with a technique, spontaneous, or promoted by other mechanisms.

	Ivs 1 60% to 70%	Ivs 2 70% to 80%	Ivs 3 80% to 90%	Ivs 4 90% to 100%
Sustaina-bility	Little self-sustained.	Relatively self-sustained.	Self-sustained.	Strong and lasting self-sustainment.
Robus-tness	Poor resilience, which can be dissipated by endogenous factors, such as lack of concentration or surprise, or by exogenous factors, such as interference of nonphysical consciousnesses.	Resilient, even when under some form of opposition from nonphysical consciousnesses or under interference of conflicting endogenous factors.	Evident resistance to energetic, emotional or mental attack (i.e., thosenic attack) from physical or nonphysical consciousnesses. Endogenous factors are supplanted by the sole occurrence of the VS.	Immunity to interferences that destabilize the VS.

Duration	Often *fugacious* (i.e., usually lasting approximately 5 seconds) or of *ordinary duration* (i.e., usually lasting between 5 to 30 seconds).	Often of *ordinary duration*, common to VSs felt by the majority of experimenters.	Often of *ordinary duration*, but tending to last longer (*long duration*; usually of approximately 30 seconds to 5 minutes).	Often of *long duration*, with greater possibility of experiencing an *extraordinary duration* (longer than 5 minutes and, in some rare cases, under special conditions, can last even hours).
Homogeneity	Less homogeneous distribution of the activation; i.e., although the resonance and respective sensation happens throughout the energetic interface, there may be more emphasis in some parts or in specific chakras.	Homogeneous distribution and complete sensation through the energetic interface (and physical, when is the case).		
Alignment	Usually, the occurrence happens in a state of alignment of the vehicles.		Occurrence, usually, in a state of small or medium non-alignment.	Triggered, usually, in a state from small to complete non-alignment.
Intensity of sensation	Sensations are more ephemeral or indistinct.	Certainty of the VS.	Experience and sensations are remarkable, unmistakable for any other experience.	Extremely intense and ostensive sensations of activation.
Vehicle	Pleasant sensations, predominantly energetic, but still mild in comparison with the more intense levels of VS.	Energetic sensations with physical perception, as the intense energy activation brings greater clarity to the perception of the physical body and of the bioenergies aggregated to it.	Clear perception of the energetic interface, with sense of the bioenergy interacting with the physical, with the possibility of also identifying some elements corresponding to the nonphysical body.	Possibility of tri-vehicular identification (physical, energetic interface and nonphysical body), allowing, in rare cases, the recognition of the "existence" of the mental body.
Duration of activation	Temporary activation, present during the VS.		Possibility of occurrence of pulsation or activation of the chakras also after the VS, even in the hours that follow the VS.	Possibility of occurrence of pulsation or activation of the chakras for many hours or even days after the VS.
Reach of effect	Weak or not far-reaching effects, with little reach of the energetic layers where old, still stagnant energies reside. Often comforting, resulting, most commonly, in the psycho-physiological relaxation (temporary psychological and emotional relaxation which takes place usually during the VS or remains for a few hours after its occurrence).	Greater-reaching effects, but generally still mediocre, not necessarily reaching deeper blockages nor producing renovation that could lead to the removal of older or more chronic connections with undesirable consciousnesses.	Intense resonance with stirring up effects and with potential to reveal undesirable energetic elements [energy blocks, nodules, or plugs (energoplugs)] not yet identified and their causes.	

Personal reaction	Possibility of (1) being dazzled in case the experimenter free from dogmatic repression or (2) enforcing self-repression in the case of the neophobic experimenter.	For the unwary experimenter, unaware of the phenomenon, the ostensive effects and sensations in the denser vehicles (energetic interface and physical) can cause fear or worry.		
Duration of results	Usually, results are more observable during the VS or right after.	Results and benefits tend to be less temporary than in the VS of intensity 1.		Results and benefits are usually more lasting and, in some cases, even permanent.
Predisposition for energetic activation	Creates a predisposition to perceive naturovibrostasis with greater ease in everyday situations.		Creates a more pronounced predisposition to experiencing the VS in the period that follows the occurrence (usually up to a few weeks).	Possibility of greater control of the activation (modulation of the attributes of the VS).
Unblocking	Usually generates ephemeral unblockings or elimination of the blocks that are recent or superficial.	More evident unblockings, with possibility of producing energy release of medium to high depth.	Possibility of releasing old energy, even those rooted in the nonphysical body, promoting large unblockings, potentially capable of bringing the experimenter to realize the informational pattern of such energies.	Possibility of releasing deep energies, even those linked to past traumas (deeper unblocking).
Energetic uncoupling	Auric uncoupling, when the energy coupling is recent or less deep.	Auric uncoupling of deeper coupling and/or pathological chronic coupling, and, in some cases, even of undesirable couplings based on affinity.	Auric uncoupling even when the coupling is with an energetically strong pathological and anticosmoethical consciousness.	
Energetic detoxification	Detoxification of foreign energies that are incompatible with the experimenter.	More complete detoxification of foreign energies, mainly those more recently assimilated.	Detoxification of foreign energies, including those whose informational pattern the experimenter resonates more with.	Complete detoxification of foreign energies that gravitate or are implanted in the experimenter's energetic interface.
Sympathetic deassimilation	Sympathetic deassimilation when the assimilation of energies is (1) recent, and/or (2) less deep, and/or (3) less resonating with the experimenter's internal conflicts or problems.	Achievement of greater deassimilations of energetic assimilations that are deeper or older.	Achievement of complete deassimilation even of deeper and older energies.	

Self-defense	Self-defense during the VS, which can also last for some time after it, usually more effective against intrusions that resonate little with the experimenter's internal disharmony.	Noticeable self-defense during the VS, with the possibility of having a longer effect, usually lasting for up to some hours after the VS.	Noticeable self-defense during the VS, with possibility of having the effect sustained for up to a few days, advancing to a more permanent self-defense.	Noticeable self-defense during and after the VS, promoting higher natural resilience to intrusions. When experienced regularly, it heightens the experimenter's level of intrusionlessness.
Deintrusion	Disconnection, either progressive or in one VS, from intrusive extraphysical consciousnesses, especially those whose presence is recent, not chronic or related to multiple lives and not connected due to affinity of personal traits. It may cause worsening by specific intruders who do not want to lose the connection.	Beginning of the elimination, or progressive deactivation, of chronic energetic coupling with pathological extraphysical consciousnesses, who do not want to lose the connection.	Elimination or progressive deactivation of chronic energetic coupling with pathological extraphysical consciousnesses who have affinity with the experimenter. Potential to also enable the experimenter to join in more complex deintrusions process.	Distancing (temporary, progressive or permanent) of consciousnesses from one's energy field, including the distancing (usually more ephemeral, but with progressive effect) of consciousnesses of rooted, chronic connection due to multiple-lives imprint, bringing the experimenter to be increasingly closer to the condition of total intrusionlessness.
Paraproprioception	Identification of the residues of more evident chakral blockages, when they still remain, and of the less evident chakral blockages, with possibility of making the root of the respective energy "clogging" perceptible.	Identification of previously non-identified energy blockages and, possibly, of their basis and causes (usually, older blockages or from deeper energetic layers).	Remarkable self-awareness of the energetic interface and its para-anatomy.	
Paraproprioception 2			Remarkable self-awareness of the coexistence of the energetic interface with the physical, with possibility of perceiving, in an intense and vivid way, the physical brain, even becoming aware of its structure (cerebral hemispheres and other functional areas).	

General parapsychism	Increase of psychic experiences, mainly those more commonly experienced by the experimenter (during or right after the VS occurrence).	Greater increase of the innate psychic abilities (catalysis of the experiences that are congruent with the individual psychic profile or tendency of the experimenter) during or right after the VS occurrence.	Expansion of psychic abilities, including those that are of types not commonly experimented by the experimenter (during or right after the VS occurrence).	
General parapsychism 2		Conducive to greater multidimensional self-awareness, with possibility that it becomes incorporated as an intrinsic condition of the experimenter.	Facilitation of healthy looseness of the nonphysical body (reduction of the severity of its "connection" to the physical).	Encouragement of the occurrence of several forms of psychic experiences at the same time.
Projectability	The accumulation of occurrences of VS of this intensity can, with time, increase the predisposition to the conscious OBE.		Favoring of the looseness of the energetic interface and nonphysical body, with possibility to promote the non-alignment and even the OBE.	Favoring of the looseness of the nonphysical body, of the non-alignment of the vehicles, and of the conscious projection, as well as predisposing of experience of series of projections in the period that follows (usually up to some weeks). This VS may also happen as a result of the projection itself.
Projectability 2			Facilitating of the increase of extraphysical lucidity and recollection of OBE experience.	
Consciential condition	Predisposing to the ephemeral intraphysical euphoria[6] (usually for a few hours).	Predisposing to the energetic springtime[7] and/or to the intraphysical euphoria of short duration (usually of some hours or up to a few days).	Possibility to produce the energetic springtime and/or to the intraphysical euphoria of medium duration (up to a few days), or, when experienced in series, of long duration.	Possibility to produce the energetic springtime and/or the intraphysical euphoria of long duration (several days), or, when experienced in series, of even higher duration.

---

[6] State of intense joy and bliss, resulting from the sense of accomplishment of an assistantial and pro-evolution task or of a relevant step in one's existential program (life task).

[7] Period in which one is with unusually loose energetic interface, resulting in positive energy effects; healthy and extremely pleasant energy condition that is experienced by an individual in profound balance and harmony, associated with a state of serene euphoria and expansion of one's energy field (aura) and extrasensory perceptions.

Consciential condition 2	"Re-tuning" with the present moment (adjustment of mental focus / attention).	Production of energetic balance, with remarkable harmonization of the chakral system.	Pacifying effect, which can last for hours or days after the occurrence.	Effect related to the expansion of lucidity and emotional balance can be identified for several days or, in exceptional cases, even weeks after the event.
Intrusionlessness	Beginning of the identification of extraphysical consciousnesses which gravitated around the experimenter but had not been noticed up to that moment.	Greater chance of identifying with clarity the extraphysical consciousnesses gravitating around the experimenter's energy field, which were previously unnoticed.	Possibility to identify extraphysical presences, including of extraphysical consciousnesses with more chronic connections to the experimenter, hitherto camouflaged.	Increased ability to abort possible intrusions before they occur.
Energetic fluidity result	Relative increment of one's free consciential energy during the VS, with possibility of being partially maintained for a short period after the occurrence.	Evident increment of one's free consciential energy during the VS, with possibility of being partially maintained for a period of up to a few days after the occurrence.	Significant increment of one's free consciential energy during the VS, with possibility of maintaining this increment partially or completely over a period of up to a few days after the occurrence. It may ultimately increase one's basal energy fluidity corresponding to a degree of the percentage of increment of free energy obtained during the VS.	Exceptional increment of one's free consciential energy during the VS, with the higher possibility of being fully kept for a period of several days after the occurrence. It may also increase one's basal energy fluidity corresponding to a degree of the percentage of increment of free energy obtained during the VS and even increase one's inherent energy fluidity.
Effects on paragenetics				More likely to produce the identification of thoughts and emotions related to past lives, and also to improve one's paragenetics, the latter being a more probable result when the VS of this level is experienced in series.

				Effects can reach one's mental body in a more direct way (i.e., producing the "liberation" of the mental body or expansion of the consciousness as well as phenomena or states derived from this condition).
Effects on mental body				

To conclude, it is pertinent to mention the obvious notion that VS and energy work are not a panacea. Experimenters with poor intellect, discernment or cosmoethics could void the permanence of the benefits obtained.

Acknowledgments

I offer my sincere gratitude to Hugo González and Iker Vázquez for their invaluable help, respectively in the creation of the electronic form and tabulation of data of the VS Survey. I also thank the IAC for authorizing use of its copyrighted illustrations (Figures 1 and 2).

Bibliography

Alegretti, Wagner (2008). An Approach to the Research of the Vibrational State through the Study of Brain Activity, Journal of Consciousness, Vol. 11, N. 42, International Academy of Consciousness (IAC), Outubro, 2008, Portugal, pp. 221-255. (Also published in this volume).

Buhlman, William (1996). *Adventures Beyond the Body: How to experience out-of-body travel*, HarperCollins, New York.

Buhlman, William (2014). *Astralinfo Suvey*, http://www.astralinfo.org/survey-results, accessed on 20 October 2014.

Crookall, Robert (1961). *The Study and Practice of Astral Projection: Analyses of Case Histories*, The Aquarian Press, London.

Crookall, Robert (1973). *Ecstasy: The Release of the Soul from the Body*, Darshana

International, Morabadad, Índia.

Fox, Oliver (1962). *Astral Projection: A Record of Out-of-the-Body Experiences*, University Books, New York.

Monroe, Robert A. (1971). *Journeys Out of the Body*, 279 p., Anchor Books, Garden City, NY.

Muldoon, Sylvan J. & Carrington, Hereward (1974). *The Projection of the Astral Body*, first edition published in 1929, Samuel Weiser, New York.

Peterson, Robert (2003). *Out-of-Body Experience: How to have them and what to expect*, 2nd edition (1st edition 1997), Harpton Roads Publishing Co, Charllotesville, EUA.

Trivellato, Nanci (2008). Measurable Attributes of the Vibrational State Technique, Journal of Consciousness, Vol. 11, No. 42, International Academy of Consciousness (IAC), Evoramonte, Portugal, Outubro 2008, pp. 163-201. (Also published in this volume).

Trivellato, Nanci (2015). *Estado Vibracional: pesquisas, técnicas e aplicações* (Vibrational State: research, techniques and applications), IAC – International Academy of Consciousness, Portugal.

Van Dusen, Wilson (1981). *The Presence of Other Worlds*, Swedenborg Foundation, NY.

Vieira, Waldo (1986). *Projeciologia: Panorama das Experiências da Consciência Fora do Corpo Humano*, edição do autor, distribuição Centro de Consciência Contínua, Brasil, pp. 327–329 e 384–385.

Vieira, Waldo (2002). *Projectiology: a Panorama of Experiences of the Consciousness outside the Human Body*, translation of the expanded and updated Portuguese version of 1999, IIPC / IAC, NY.

Note: This article was previously published in: Journal of Consciousness 18, No. 59, 2015, p. 337.

AUTORICERCA

The vibrational state: a novel neurophysiological state?

Rodrigo Montenegro

Issue 20
Year 2020
Pages 191-231

Abstract

This article presents a theoretical assessment specific to the field of consciousness research, more precisely associated with the study of the Vibrational State (VS) and accompanying states, such as the Out-of-Body Experience (OBE), the former currently being an under-researched state of consciousness when it comes to its neurophysiological correlates. It aims at informally presenting neurophysiological theoretical assumptions gathered over the years studying the VS as well as the OBEs through empirical observations, although briefly presenting current ongoing scientific experimental electroencephalographic research into the VS as well as prospective research and scientifically based neuroscientific theories in preparation, which taken together, may provide recommendations for future research associated with such states. We present succinct preliminary research data of the VS intending to postulate insight into the current limitations of VS studies. The purpose lay forward is that of elaborating a critical framework to technically define the VS as a peculiar neurophysiological state of consciousness, not only as an experiential one. Although this essay is merely speculative and mostly theoretical, it is inferentially based on experimental data, current neuroscience and sleep research knowledge. Nevertheless, it does not pretend to have reached the level of a structured hypothesis based on reproducible scientific evidences at any point, when those are presented. Some of the presented assumptions are certainly disputable and based on inductive, experiential-based, reasoning or limited data analysis. Nonetheless, the insights, taken forward, may provide research implications for the neurophysiological models accompanying future VS and OBEs research, including considerations regarding Isolated Sleep Paralysis (ISP), Lucid Dreaming (LD) and other correlated states. We further assume that such states are associated with the brainstem region, have possible common neurophysiological substrates, even though they are phenomenologically different states of consciousness, presenting state-dependent markers.

AutoRicerca, Issue 20, Year 2020, Pages 191-231

1 Introduction

> *It would be most satisfactory of all if physics and psyche could be seen as complementary aspects of the same reality. To us [modern scientists], unlike Kepler and Fludd, the only acceptable point of view appears to be one that recognizes both sides of real-ity — the quantitative and the qualitative, the physical and the psychical — as compatible with each other, and can embrace them simultaneously. – Wolfgang Pauli*

The Vibrational State (VS) is defined as a non-ordinary state of consciousness spontaneously perceived during sleep or voluntarily induced through specific meditative techniques, as well as by entheogens experiences. The experience is often described by the proprioceptive perception of vibrations, such as "tingling" or "electrical" sensations. However, although such sensations are often mistakenly described as the VS, as a state of consciousness the VS is rather defined by the ending condition characteristic of such vibrations: final state universally designated as a powerful and intense sensation of being like a "dynamo" (Vieira, 1981, p. 128) or being in a profound, penetrating state of "resonance" (Vieira, 1986; Alegretti, 2008, p. 234).

Albeit the word *state* is employed to describe the subjective set of the VS subjective phenomenology, little is known about the neurophysiology of such state, despite its universal and historical description, at least since Socrates. Historically, in Phaedrus, which portrays Socrates' dialogues (c. 470–399 BC) with Pythagorean philosophers, Plato describes the experience of the throbbing of the entire soul (Phaedrus 251-a-e). A state of vibration brought up by an entheogen experience arising from Eleusinian practices (Hamilton, 1973; Rinella, 2000; Wasson, Hofmann and Ruck, 2008) equally described in modern DMT experiences (VS-DMT) (Callaway et al., 1999; Strassman, 2001; Tittarelli et al., 2015; Hamill et al., 2019) clearly suggestive of comparable VS phenomenology.

The experience described by Plato at the end of the initiation, leading to the perception of "pure light" (Phaedrus 250-c) is reminiscent of Buddhist's consciousness states concepts (Bhardo, in Tibetan), following experimental observation made since the 10th century by the Indian tantric teacher Nāropā. Such meditative practices described non-ordinary meditative states mostly associated with sleep, whose purpose was preparation for death. The *bhardo of dharmata*, for example, is thought to be experienced by maintaining the perception of "clear light" at the sleep onset (SO) and presents similar experiential vibration phenomenology, as do other practices such as the Yoga of Tummo, Pho-wa meditation and Kundalini Yoga (Evans-Wentz, 1958; Gardner et al., 1993). Practiced through the stimulation of "energy flows" or "wind flows" (Gardner et al., 1993, p. 104; Rinpoche, 2012) and driven through deliberate "pulsations" of subtle energies directed bottom-up (Gardner et al., 1993, p. 103; Rinpoche, 2012) through the meridians of the body, such practices were accompanying processes of ego-dissolution and led to the transference of consciousness out-of-the-body. Although such meditations practices have often utilized specific terminologies to describe such processes, the associated experienced are clearly allusive of similar but contemporary VS-induction methodologies, undeniably illustrating the meditation to prime the experience of "a body of light" or "illusory body" (Rinpoche, 2012). An incidence equally prevailing in modern VS accounts inducing Out-of-Body Experiences (VS-OBEs) through the projection of the astral body (Montenegro, 2015).

Although Western traditions provided phenomenological insight into sleep phenomena, since Plato described dreams of foretelling (Oneiroi, in greek) (Theochari, A., 2008) in "Republic" and Aristotle "De Insomnii" studied the ability of Lucid Dreaming (LD) (De Koninck, 2012) – a condition where the dreamer expresses increased awareness –, no description of the VS appears to be illustrated in Western societies until 1744. The notion seems to have been first brought by Swedenborg's journal of dreams and spiritual experiences describing the VS as "powerful tremor, from the head and over the whole body" (Swedenborg, 1918, p. 26), a condition re-counted to often progress to OBEs. However, despite the reference, little historical accounts of the VS are found until the twentieth century. A phenomenological standstill which might have been caused by the religious censorship following the otherwise notable work of theologian Quintus Septimius Florens Tertullianus (c. 155 – c. 240 AD) on sleep states, acknowledging dreams as a doorway to demonic influence (De Koninck, 2012).

The standstill might have been equally brought up by traditional notions of sleep, dating back from Hippocrates, which by promoting physiological explanations of the sleep phenomena (De Koninck, 2012) also led to a certain reductionism, in the name of scientific progress. For instance, Isolated Sleep Paralysis (ISP) – a healthy condition of sleep atonia, leading to the impossibility to move in sleep when we wake up, – was first referred in 1644 as an Incubus Night-Mare (Kompanje, 2008), without however being accompanied by descriptions of interoceptive vibrations (ISP-VS), until Everett first described them in 1963, as a reported condition of "tingling sensation over the body" (Everett, 1963, p. 284). Similarly, until the nineteenth century, most sleep studies, such as the work of Louis Ferdinand Alfred Maury (1865), who discovered the hypnagogic state at sleep onset (SO) – where the VS seem to be occasionally self-reported – does not seem to refer either to VS-like perceptions, despite lexicon limitations.[1]

Notwithstanding the development of the experimental approach at the time, research was mostly focused on apprehending sleep essentially from a physiological standpoint (Morin and Espie, 2012; Maury, 2013) and if such studies made valuable psychological observations, they remained mostly ascribed to such pursuits. Consequently, much less attention may have been given to phenomena such as the VS, which may have been simply dismissed as phantasmagorical hallucinations, as they are still today.

Self-evidently, the same condition might have subdued some of Sigmund Freud's conclusions on dreams. Although Freud recognized that faculties of the mind, such as the intellect, remained intact in dreams states (Freud, 2010, p. 89), specific capacities of consciousness in sleep, like the lucid dreaming (LD) states reported by Saint-Denys (1867) (Hervey de Saint-Denys, 2005) as the ability to direct dreams, were downplayed as preconscious wish (Freud,

---

[1] The term "vibration" was mostly used at the time to refer to nerve activation, e.g., "cerebral vibration."

1965, p. 571). However, their analysis could have led to the description in such states of "electrical sensations," specific to the VS sensations reported in LD (LD-VS) (Levitan et al., 1999; Waggoner, 2009), with an acknowledged Rapid Eye Movement (REM) state prevalence, although the statistical relevance of such states in LD has received little formal quantification.

Descriptions of the VS started to flourish at the beginning of the twentieth century with the upsurge in OBE reports led by classical authors such as Muldoon and Carrington (1929); see also (Montenegro, 2015). A condition that might have culminated with a wider divulgation of the VS, as a state that can be induced, by Monroe in the 70s (Monroe, 1971). However, it is only in the late twentieth century that prominent researchers of institutions such as the American Journal of Psychiatry started to compile the statistical prevalence of the vibrations, which a randomized OBE survey reported to be perceived by **52%** of participants ($n = 339$), and by **38%** for VS-OBEs (Twemlow, Gabbard and Jones, 1982).

Nevertheless, whilst non-randomized surveys continued to report the statistical prevalence of vibrations to be between **30%** ($n = 1115$) (Alegretti and Trivellato, 1999) and **56%** (Buhlman, 2001), the neurophysiology of the VS in sleep states, to date, has never received the scientific attention it merits. Besides its distinct typology, the VS has, arguably, remained mainly associated with OBEs, and because of this probably remained mostly downplayed as a hallucinatory phenomenon (Cheyne, 2010) ascribed to psychiatric disorders of perceptions (Montenegro, 2015).

Finally, if modern investigations on states of consciousness have been polarized by a research mainly focused on neurophysiological correlates and cortical determinants of such states, and not of their associated phenomenology (Aru et al., 2019), it is somewhat surprising to see a similar reductionism in empirical studies of the VS. The VS is known in consciousness investigation circles for more than half of a century, with empirical studies reporting an increasingly refined characterization of such state. Such studies, however, remained impervious to a research of its neurophysiological correlates, despite acknowledging their importance. Without a doubt, until today, little is known about the VS from a neurophysiological standpoint, which would undoubtedly provide a deeper understanding of the conditions that would favor its triggering and could, as such, increment motivation for its induction.

Albeit preliminary data on the VS was provided by Alegretti (2008) and Rodrigues Pinheiro (2013), the current scope of the research related to the induction of the VS during the waking state (W-VS) does not meet those criteria that would allow the VS to be defined as a specific and novel neurophysiological state, taking into account modern concepts of state determination,[2] considering that the VS is spontaneously perceived in LD-VS and ISP-VS sleep states. A condition, in view of the assumption proposing the VS as a mechanism to explain OBEs (Vieira, 1986, p. 109),[3] considered of fundamental importance to allow for state differentiation.

The focus of this essay is precisely on the notion of state differentiation. More precisely, its objective is to delineate, at least theoretically, the experimental research needed to determine the VS as a novel and testable neurophysiological distinctive state. However, a review of known neuroimaging methodologies limitations or current neurophysiological theories that explain OBEs associated with this subject, specifically nosological, as well as notions related to their ontological reality, or its acknowledged relation with VS-DMT, would go beyond the scope and objective of this article. Also, those multidimensional characteristics that could prevail and may prompt such states are equally left aside of the present epilogistic analysis, to focus on the identification of possible neural correlates of consciousness in such states, although they could involve important causal attributes.

---

[2] These are referred to as the neurophysiological determining properties (neurophysiological markers) of each known stage of sleep: Wakefulness (W), R-stage (previously referred as REM), N-stages (3 Non-Rapid Eye Movement phases – N1, N2 and N3 stages).

[3] The relationship correlating subtle energies (e.g., chi, prana, orgone, etc.) to the induction of psychic states were hypothesised since the nineteenth century by many empirical researchers, such as De Rochas, Durville and Carrington (De Rochas D'Anglun, 1895; Durville, 1909; Carrington, 1921), and the relationship to OBE was incipient in research led by Crookall since 1960 (Crookall, 1978, 1980, 1998).

2 Preliminary analysis of the physiological characteristics of the vibrational state

What do we know of the VS as a neurophysiological state? If a comprehensive methodology establishing the modus operandi to achieve the VS has been published by Trivellato (2008; 2015), the knowledge of the VS as a neurophysiological state still remains in infancy. Sparse Electroencephalographic (EEG) VS data from an initial 2-session experiment was initially presented by Alegretti (2008) but did not offer enough technical information to provide a well-grounded critical appraisal. If the author acknowledged that the study lacked the desired scientific rigor, several observations of unusual brain patterns certainly lack a more detailed technical explanation to infer about the observed conditions.

For example, the referred "synchronization of several brain circuits" (Alegretti, 2008, p. 248), could refer to a diffuse polymorphic Theta activity associated with deeper drowsiness states (Stern and Engel, 2013). However, drowsiness was not reported.

The condition of "cerebral arrhythmia" (Alegretti, 2008, p. 248) remains an equally unclear depiction to provide any sensible specification, despite the author's communication attempts for clarification with Alegretti. If the neural terminology could be referring to a condition of partial seizure, related to asymptomatic ictal bradycardia episodes seen in seizure without syncope (Almansori, Ljaz and Ahmed, 2006; Bartlam and Mohanraj, 2016), they are very improbable and uncharacteristic of VS phenomenology.

The second study of Alegretti, mentioned in the same article (Alegretti, 2008), despite describing similar patterns, and more specifically the presence of high-frequency waves, does not offer technical data to allow in-depth analysis either. Further unpublished studies of Alegretti, presented in conferences, have been mainly concerned with energy detection, not neurophysiological brain activity, or the neurophysiological correlates of such state.

Rodrigues Pinheiro (2013) completed a subsequent Master

dissertation study on the subject. The study specifically focused on the electrophysiological correlates and investigated 25 subjects, including 15 experienced participants with ten years or more of experience inducing the VS, as well as ten controls. Results analyzed 20 bipolar cortical channels of subjects and control attempting to reach the VS with an general limited significance analysis level of $p < 0.05$ for intergroup differences. An overall Gamma burst of up to 80 Hz ($p < 0.01$) was however observed in the experienced population and suggested a broader cortical Gamma during the VS (Rodrigues Pinheiro, 2013), although such hypothetical conjecture remains to be verified due to the limited spatial cortical reach of the 20 EEG channel setting to provide certainty of the fact. It was not observed in the research led by this author (Figure 1).

The relation to sleep stages, critical to state differentiation, was briefly mentioned. The study indicated the occurrence of light sleep with increased Theta waves (in **50%** of the **30** s EEG segment) and the occurrence of sleep with the apparition of vertex waves **10%** of the time during VS-induction (I-VS) by experienced subjects; however no further analysis was given. The condition of light sleep is in itself, otherwise atypical, taken in consideration the known arousal effect of Gamma waves and was not reported during I-VS in the W-stages in our study.

Whereas the report is consistent with current neurophysiological research, showing an increase in Gamma above the indicated threshold during VS, in a **128** high-density EEG channel investigation led by Montenegro (*in preparation*), a comparative study suggests differences in cortical areas with the referred schoolwork which might be prompted by differential I-VS techniques. More so, current research equally presents data showing a significant overall Delta increase concomitant to Gamma during I-VS (Figure 1), not reported by the previous study by Rodrigues Pinheiro (2013). This is a neurophysiological correlate that may be ascribed to an increased focused state (Montenegro, 2020a).

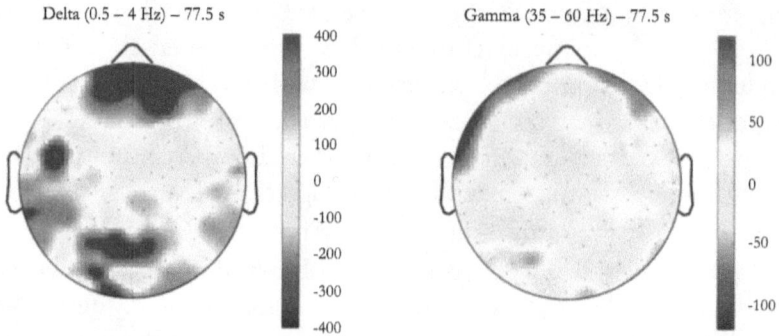

Delta (0.5 – 4 Hz) – 77.5 s   Gamma (35 – 60 Hz) – 77.5 s

Figure 1 128 high-density EEG showing concomitant Gamma and Delta with percentage increased VS (unprocessed data) during one VS-session (in color in the pdf version of the article).

If the research in preparation delves deeper into the neurophysiological characteristics of the VS, providing a theoretical framework to suggest a specific neural signature (Figure 2) suggestive of prospective Event-Related Potential (ERP) or waveform, with specific Phase Amplitude Coupling (PAC) and Phase Locking Values (PLV),[4] consequently acknowledging the state to be mediated by specific neurotransmitters, the research still does not meet the criteria to ascertain state differentiation. A condition more specifically formulated subsequently.

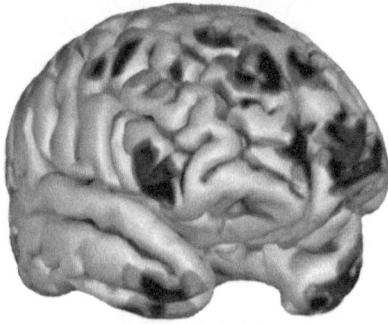

Figure 2 Neuroimaging of cortical areas associated with the VS during VS-induction in a specific VS-session (in color in the pdf version of the article).

---

[4] PLV are EEG-data statistical analysis of task-induced changes in neural synchronization allowing measuring functional connectivity. PAC, on the other hand, measures the coupling of band wave synchrony.

Whilst the referred research indicates the VS to be an *objective* non-ordinary state with specific and novel properties, there was a preexisting need to establish the VS within the better-defined boundaries of the waking state. Initially, preventing LD-VS and ISP-VS state comparison. A limitation inherently steaming, as suggested by Abraham (2010), from the fact that I-VS are led by top-down cognitive correlates, which consequently need to be better dissociated from Spontaneous-VS (S-VS).

Furthermore, no specific methodology has been proposed to allow a better distinction between S-VS seen in the different phases of sleep. If the research proposed by (Alegretti, 2008, p. 244), suggesting the comparison of EEG results of experienced subjects, would permit for its possible spontaneous generation in deep levels of relaxation, the experimental data gathered would suffer from the proximity of the recording to the waking state. As such, if the experimental setting may provide consequent data, at this stage of the research, it would not allow for the analytical comparison between I-VS and S-VS in R or N-states, despite the experimental necessity referred above. A condition *sine qua non* to allow for further state differentiation beyond the known phenomenological similarities of such states, so as to obtain the necessary distinction with respect to their associated overlapping states (see the analysis in the section below).

Frontal low current electrical stimulation induction of a gamma activity might be better suited to the differential objective permitting an increased lucidity in R and N-states and consequent S-VS in predisposed subjects. A combined EEG research protocol including transcranial Alternate Current stimulation (tACs) and stage detection algorithms may allow for a programmed neurostimulation targeting cortical areas during certain neurophysiological threshold of sleep, allowing the stimulation of brain areas, such as the frontotemporal cortex, to stimulate the awareness of subjects and allow them to attempt at inducing S-VS or indeed I-VS. A condition not devoid of technical complexities and limitations, but certainly possible.[5]

---

[5] The electrical potential of such stimulation can be subtracted from EEG recording, allowing for a more objective data analysis. The neuromodulation could potentially occur during LD states or induced ISP, where Alpha lucidity is reported (Takeuchi, 1992), and confer neurophysiological state similitudes.

Nevertheless, another more pressing concern seems even more critical to address, in structuring a larger-scale and more in-depth investigation of the VS, allowing for the proposed state differentiation – one that would delineate a scientific neurophysiological theory of the VS as an original neurophysiological state. Undeniably, the facts suggest that most subjects lack the capacity of identifying an objective reference to measure their subjective states, or are inadequately trained to induce the VS. Although the VS is deemed to be easier to learn than OBEs, constituting theoretically a rather replicable phenomenon, empirical and subjective analysis of the VS suggests that **95.7%** of subjects were unable to reach the VS without long-term guided training (Trivellato, 2014). Moreover, **30%** of studied subjects thought to have achieved the VS without objective indications of the fact. The lack of experiential understanding of the phenomenology associated with the VS is a condition that might further compromise the statistical analysis of any replication study, although current technological trends of consumer EEG might help experiencers to characterize better the state with objectivity, albeit they might also be limited by the technological limitations of such devices to measure the VS.[6]

3 Neurophysiological markers of state determination

Defining the VS as an original state requires it to be unambiguously differentiated from acknowledged neurophysiological stages of sleep, i.e., distinguishing ISP-VS, LD-VS and S-VS, from W-VS. This can be achieved by understanding the differential neurophysiological markers of state determination associated with

---

[6] Analysis of the current market indicates most consumer grade headbands present technological limitations to measure high Gamma wave (e.g., above < 64 Hz) as they are reported. A condition led by the Nyquist frequency sampling theory, which defines EEG recording sampling threshold need to have sampling recording rate of at least double the frequency (e.g., 64 Hz) to allow accurate recording of half the targeted frequency (e.g., 32 Hz).

such states.

Sleep research has conventionally recognized three different states of being: W (wake), R (REM sleep) and N (non-REM sleep) stages. They consist of different states of consciousness or unconsciousness, with specific neurophysiological markers (biological sates) and associated variables characterizing each stage. However, although such states may express differences from a neurophysiological standpoint, they may present in some circumstances, relatively "similar" states of consciousness, as exemplified by dream activity in R and N stages (Suzuki, 2004; Manni, 2005). On the other hand, such states may be different states of consciousness, as exemplified by the differences in the W and R states (excluding LD), but nevertheless express neurophysiological similarities (Mahowald and Schenck, 1999; Fitzgerald, Gruener and Mtui, 2012). This is one of the reasons why the R-stage was also termed *paradoxical sleep*, as it presents EEG wave markers that are seen during the W-stage (e.g., Beta waves), although it is mostly a Theta predominant state.

The determination of the W, R and N stages is carried out using various criteria, and technical instrumentation measuring electrographic activity through a plethora of multi-parametric polysomnographic (PSG) equipment, including EEG, electrooculogram (EOG), electromyogram (EMG), electrocardiogram (ECG) pulse-oximetry, and airflow/thermistor, allowing for the determination of the properties of each state. However, despite this technological predominance, behavioral assessments (e.g., conditions of the eyes, movements, reactivity to the environment) have remained necessary and are equally useful for state determination (Mahowald and Schenck, 1999), although an growing trend in using algorithms' scoring, which have reached high accuracy level regarding their agreement with manual human scoring.[7]

Defining the standard W, N and R stages is generally achieved by recording and analyzing voltage fluctuation differentials through electroencephalography (EEG) recording. EEG frequencies (Table

---

[7] Such software has high levels of validity only when they present an adequate sensitivity, so as to rule out other stages of sleep generally accessed by *F*-measure statistical analysis to consolidate their precision.

1) are combined with concurrent monitoring of muscle tone, measured by electromyogram (EMG) signals, and eye movement recording, ascertained by electrooculogram (EOG) signals.

Band	Frequency (Hz)
Delta (δ)	> 3.5
Theta (θ)	4-7.5
Alpha (α)	8-13
Beta (β)	14-30
Gamma (γ)	< 30

Table 1 Frequency bands as indicated in Niedermeyer's electroencephalography (Schomer and Lopes da Silva, 2011).

The natural state of wakefulness is characterized by Low Voltage Fast Activity (LVFA) and cortically synchronized EEG patterns for higher frequency ranges, evolving to Theta increment in the frontal cortex, more specifically with increased sleep drive. Wake frequencies range from Beta to Gamma during intense mental activity, while relaxed wakefulness states have predominant Alpha waves, depending on behavioral input. In general, the W-stage with eye closed is characterized by a dominant Alpha rhythm, most evident on the occipital EEG area. The overall W-stage expresses otherwise the presence of diverse patterns of concomitant non-exclusive brainwaves (Table 1), even for a same functional area.

Sleep, on the other hand, is more characterized by large-amplitude slower wave frequencies (N-stages), which is an overall EEG condition led by the reduced activity of the Ascending Reticular Activating System (ARAS). N-stages progress from light sleep (N1) to deep sleep (N3). N1-stages are characterized by low amplitude, mixed frequency EEG (**4-7.5 Hz**) with central EEG vertex sharp waves (V waves) (**< 0.5** s duration), and present Slow Eye Movements (SEMs). N2-stages express fast bursts (**≥ 0.5** s) of **11-16 Hz** activity of sleep spindles, most apparent on central EEG, K complex waves, mostly manifest on frontal EEG. Slow waves (**0.5-2 Hz**) Delta activity characterizes N3-stages of deep sleep in at least **20%** of the EEG activity with no eye movements, with **≥ 75 μV** in amplitude in frontal EEG and sleep spindles (**≥ 6 s**).

R-stages, similarly, considered as light sleep, are characterized by bursts of Rapid Eye Movements (REMs), low amplitude EMG (atonia), sawtooth waves, most evident on central EEG, often pre-ceding bursts of REMs and transient muscle activity characterized by phasic twitches. Figure 3 provides a hypnogram with typical brainwave patterns associated with each stage.

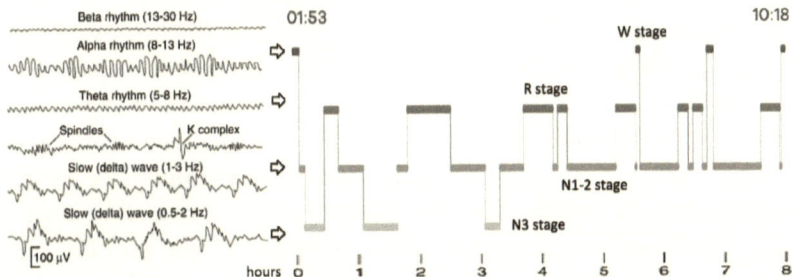

Figure 3 An adult sleep of 8-hours, with 21 minutes in W-stage (**4%**) through the night and typical EEG patterns associated with each stage (top picture), although associated with an increased R-stage pattern (4:16). N1-stages commonly express Low Amplitude Mixed Frequency (LAMF) **4-7** Hz EEG with Vertex Sharp waves (V waves) and N2-stages record sleep spindles with fast bursts (> **0.5** s) of **11-16** Hz activity and K-Complex. While N3-stages shows slow waves in > **20%** of the epoch (0.5-2 Hz Delta activity). Spindles can persist into N3-stages. R-stages have bursts of Rapid Eye Movements (REM) with low amplitude EMG (atonia) and transient muscle activity (Phasic twitches) and might express similar EEG patterns to the W-stage. Here the average heart rate was **56** beats per minutes and an average breathing rate of 14 breaths per minutes, with a total of **24** position changes during the sleep duration.

Without entering too deep into the complexities of the conditions of state determination, it is essential to denote that such electrophysiological patterns induce electrical properties changes in the function of signal integration, modulating ions channels potentials and consequently determining synaptic and neurochemical properties of neurons (e.g., inhibitory or excitatory). The neurophysiological condition of state determination is therefore multifaceted, with a state definition that goes well beyond the EEG electrophysiological assessments. A complex variety of factors associated with state determinants are expressed in a wide variety of neurophysiological parameters at play, to name a few: field potentials, postsynaptic currents, state of synaptic distribution, neural anatomy, neural properties, resting potential, receptor gating, equilibrium. These parameters should be in principle taken into consideration in the analysis of the VS, if one wants to be able to explain its neural modulation.

Additionally, while a diverse set of neurotransmitters and neuromodulators modulate the prevailing and cyclic properties of each state, they may do so by operating on the same neural networks, remaining anatomically "interpenetrated" in their neuroanatomical functions (Hobson, Lydic and Baghdoyan, 1986, p. 371), but nevertheless expressing different states of consciousness. That neural condition is reflected in the expression of the pontomedullary reticular formation (RF) and reticulospinal neurons (RSNs) in the brainstem region, which allow for motor suppression during the R-stages (sleep atonia), while allowing for behavioral motor control in W-states and, as such, are a multimodal and state-dependent network (Takakusaki et al., 2015; Brownstone and Chopek, 2018).

Finally, state determination, as indicated, is defined by the distinctive determination of specific neurotransmitters delineating the prevailing properties of each network. If, as exemplified, the same neuroanatomical networks may convey different prevailing state-properties, with different states of consciousness, to an equal extent, the same neurotransmitters, such as 5-hydroxytryptamine receptors (5-HT), and others, play equal roles in W-states as they do in N and R-stages.

Wake/Sleep stage induction are mostly modulated by either Orexin, Histamine (H), 5-HT and noradrenaline (NA) for the waking promoting nucleus of the hypothalamus, while Gamma-Aminobutyric Acid (GABA) mostly modulates sleep states associated with the Ventro Lateral Preoptic nucleus (VLPO) (Figure 5).

Last but not least, state-dependent neurophysiological properties are likewise predisposed by genetic factors or transcriptome genes that drive the property of each state (genetic characteristics of state determination). Undeniably, functional physiological mechanisms accompanying healthy circadian sleep are associated with Transcriptional-Translational Feedback Loops (TTFLs) that initiate the wake/sleep cycle, a condition mainly regulated by the genetic function of three genes: Period (Per1, Per2, and Per3), Clock (Clk) and Cryptochrome (Cry1 and Cry2) genes), expressed in the neurons of the Suprachiasmatic Nuclei (SCN) (Figure 5). The circadian predisposition is further influenced by waking states and regulated by different neurotransmitters, such as Adenosine A1 receptors levels affecting sleep propensity, and is a condition which, amongst many others, allows for healthy circadian cycles to occur. As such, even though no evidence currently suggests a genetically causal drive for the VS or VS-OBE, we may assume that, as research progresses, one may discover similar genetic factors predisposing such states. Such a genetic profile could influence the spontaneous generation of the VS, as they predispose Gamma wave generation or Slow Wave Sleep (SWS) propensity in N-stages (Wulff et al., 2010) and ISP, the latter being influenced by Per2 gene polymorphism (Denis et al., 2015), a genetic outline specific to a circadian dysregulation prevalence (a condition known to lead to ISP proneness).

As mentioned, the above neurophysiological notions are relevant to determine and differentiate the neurophysiological properties of the VS, if it is to be defined as a novel neurophysiological state from a scientific standpoint. Two aspects are essential to consider in such respect. Firstly, if LD states have relatively better-known neurophysiology,[8] ISP states lack neurophysiological consensus and more specific properties to provide a well-established state differentiation and consequent distinction with the VS. Secondly, there are the complexities associated with transitional states, defined as overlapping, entangled neurophysiological states markers of relative transitional nature, where the VS is reported to occur, as explained below.

4 Transitional sleep states

The concept of state, as a condition of the mind, has considerably evolved since the discovery of REM (Rapid Eye Movement) sleep in 1952, by Aserinsky and Kleitman (Morin and Espie, 2012). Current research has grown beyond the limits set by the three known

---

[8] Although much more research has been carried for LD states, LD studies have suffered from limited samples with limited spatial EEG resolution and as such, have limited statistical power for determining the prevailing neurophysiology of LD, which is, to a certain extent, still in debate in terms of functional areas.

prevailing W, N1-3 and R-stages, to equally consider overlapping transitional states, some of which express clinical state dissociations. In fact, if neurological states' determination usually has prevailing conditions, sleep laboratory research has led to the observation that neurophysiologic markers may illustrate a condition of the interlocking of neurophysiological states into one another. Such condition is the cause of the extensive array of state dissociations (Mahowald and Schenck, 1999; Mahowald and Schenck, 2005) and parasomnias disorders such as confusional arousal, somnambulism (sleepwalking), sexsomnia and REM Behavior Disorder (RBD), among others. In the case of overlapping states, the otherwise normal and healthy transition between each state, gradually allows multiple state-determining markers to appear in concomitance (Terzano, Parrino and Spaggiari, 1988; Mahowald and Schenck, 1999), which are often inductive of clinical dissociative variations.

In wakefulness variations, such as in narcoleptic cases, dissociative states arise from wakefulness and are accompanied by sudden sleep and the experience of REM intrusion (Overlapping State-1; see Figure 4). The condition of cataplexy in Narcolepsy is the sudden transient episode of muscle tonus loss considered to be triggered by the sudden intrusion of the R-stage into wakefulness. A condition that has often been misdiagnosed as schizophrenia in the past. Hypnagogic or hypnopompic mental activity or hallucinations are considered dream/wake overlapping states variations (Overlapping State-2; see Figure 4) (Mahowald and Schenck, 1999; Manni, 2005). Somnambulism or confusional arousals, equally known as sleep "drunkenness," are N-stages sleep prevailing variations, generally experienced without lucidity, even though amnesia is not always complete (Overlapping State-2 variation; see Figure 4). R-stages prevailing variations are in turn perceived in drug intoxication or withdrawal states, but are more often documented in RBD, where the loss of muscle atonia occurring during the R-stage sleep leads patients to act their dreams.

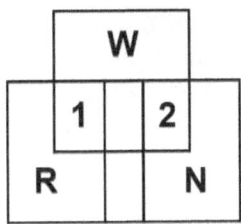

Figure 4 Overlapping neurophysiological states 1-2.

Such characteristics of interlocking states are relevant to the study of the VS state differentiation, explicitly so in LD-VS and ISP-VS. A reflection of the fact that ISP is mostly understood as the persistence of R-stage atonia into the W-stages (Overlapping State-1 variation; see Figure 4), although it has been suggested that ISP can be recorded at predormital SO (Takeuchi et al., 1992; Girard and Cheyne, 2006) (Overlapping State-2 with N-stage variation; see Figure 4). A condition, which in Recurrent-ISP (RISP), considered a pathological variation of ISP, may last for several minutes and occur several times through the night, unlike ISP. LD experiences are equally considered as an R-stage sleep prevailing state, with a mixed W-R-state (Mahowald and Schenck, 1999) (Overlapping State-1 variation; see Figure 4) allowing for the activation of functional areas such as the Dorso Lateral Pre Frontal Cortex (DLPFC) associated with decision-making, working memory and social cognition.[9]

Taken the relatively short forty years history of modern sleep research, we may assume that other transitional and overlapping states could still be discovered. Such might be the case for the state of *projective catalepsy*.[10] The cataleptic trance was referred to as a specific "state of motionless, of sustained immobility, without or

---

[9] REM genesis is correlated with hypothalamic and amygdala activity associated with the transient inhibition of activity of the DLPFC, which explains the lack of emotional context and rationality during dreams. The dreamer becomes unable to refer to relevant information or questioning. Although such neurophysiological conditions of inhibition stimulate the expression of emotions and their spontaneous regulation, LD states, through the DLPFC activation, seem to promote such regulation in a much more evolutionary fashion.

[10] The term *projective catalepsy* was initially used in an attempt to distinguish it from ISP terminology, considering that ISP was initially perceived as a pathological sleep phenomenon.

without clouding of the sensorium," with patients exhibiting the appearance of sleep (Kinnier Wilson, 1928, p. 90).[11] The state was proposed as a transient state of partial consciousness disconnection and equally as a functional mechanism to explain S-VS predisposition, leading in some instances to OBEs (Vieira, 1986).

Swedenborg acknowledged the above view already in 1758, as an experiential state of "insensibility" (Swedenborg, 1884, p. 354-355), or *deep torpidity*, allowing OBEs, with similar assumptions early laid by Lancelin (1910) and Muldoon and Carrington (1929). Cornillier (1927) discussed the mechanism at the time of death. Reported by Eliade as "cataleptic consciousness" (Salley, 1982), many OBE experiencers have experimentally induced such state at the SO, as a bridge to OBEs (Lancelin, 1910; Muldoon and Carrington, 1929; Fox, 1962; Monroe, 1971; Vieira, 1981; Rogo, 1986; Montenegro, 2015), thus constituting *voluntary induced* pre-OBE-Catalepsy (OBE-C) as a possible mechanism to explain OBEs – the so-called mind-awake body-asleep *doorway of perception*.

Embryonic explanations related to such state did not lead, however, to the development of a genuine scientific theory, although the causal mechanism of insensitivity was briefly associated with the medulla oblongata and OBEs to the rhomboid fossa as from 1958 (Vieira and Xavier, p. 120, 129). Whereas these authors did not provide information to delineate the neurophysiological mechanisms of the inferred association, the underlying idea here, within the limited anatomical functions of the discussed areas, presumably signifies that mechanisms of motor inhibition at the SO, or naturally occurring during sleep, although not suggested by these authors, would predispose to OBE experiences. A more contemporary terminology would instead hypothesize the inhibition to be related to the mesopontine tegmentum and the medullary reticulospinal tract in the pontomedullary Reticular Formation (RF) (Figure 5), within the networks of the brainstem, which would exhibit specific state-dependent characteristics, mediating transient pre-OBE-states (Montenegro, 2020). This is an inhibitory motor hypothesis which

---

[11] The term catalepsy was at the turn of the last century interchangeable with the term cataplexy, or the sudden loss of muscle tone in the waking state, although the term catalepsy prevailed as a specific state of insensibility.

has been defended by the present author since 2005, as a naturalistic mechanism to explain the universally experienced prompting of OBEs in sleep (Montenegro, 2015).

The above consideration is of fundamental importance, as it does not only reflect the discussed hypothetical assumption proposing the VS as a mechanism to explain the OBEs, but the idea, further stipulated, that the inhibition of the functional areas, specifically associated with the pons (Figure 5), are supplementary to the neurophysiological mechanisms of LD-VS, ISP-VS, and VS-like sensations, although state-dependent, *transitional states variations*, leading to OBEs. As such, they have functionally *analogous* neurophysiological characteristics of motor inhibition and atonia, as seen in R-stages or ISP-states, although here hypothesized to be state-dependent.

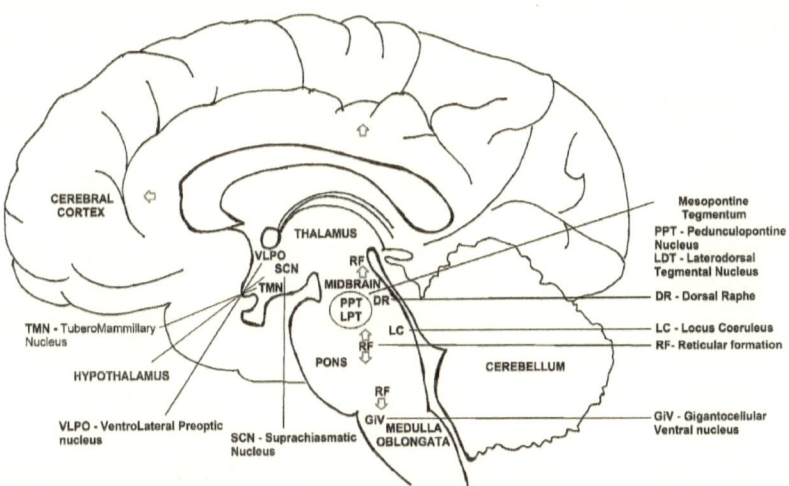

Figure 5 Sagittal sections of the brainstem composed of a complex network of interconnected regions from the thalamus/hypothalamus downward, forming the Midbrain, the Pons, and the Medulla Oblongata. Such regions are further comprised of different brain nuclei such as the Ascending Reticular Activating System (ARAS), reaching across all three areas with pathways to the Cerebral cortex (arrows) and forming a complex functional network that promote and control the sleep-wake cycles driven by the Suprachiasmatic Nuclei (SCN) transcriptome genes. The schematics also indicate other regions mentioned explicitly in this article.

More specifically, based on an unpublished investigation, briefly outlined below, we can presume that R-stages' variations could express characteristics of Sleep Onset REM Periods (SOREMPs), a condition where REM is uncharacteristically perceived at the SO. Although ISP-like variations, where similar mechanisms of R-stages atonia are at play, would express the same common neural denominators, induced OBE-C at the SO would be prompted through the voluntary inhibition of ARAS systems, as delineated subsequently.

R-stages variations would possibly be prompted through cholinergic hyper-activation (REM-ON) states triggered at the SO, associated with an inhibition of REM/OFF Locus Coeruleus (LC) nuclei, which within the mesopontine subcoeruleus (SubC) induces a very low amplitude EMG muscle tone (R-Stage atonia), modulated by GABAergic motoneurons of medullary pathway of the Gigantocellular Ventral nucleus (GiV) (Figure 5). However, while R-stages are usually modulated during the night, the induced R-stage REM-ON variation would be expressed at the Sleep Onset REM (SOREMPs), without loss of lucidity. This is a sleep state that seem to correlate to an OBE study conducted by Tart (1968), with a gifted subject, but also confirmed, to a certain extent, by an unpublished OBE-EEG preliminary investigation conducted by this author.[12] The state suggest a condition that is not unlike Wake Induced Lucid Dreaming (WILD) induction, albeit repetitively realized at SO and not throughout a night waking schedule, in the second half of sleep, where REM has a higher prevalence. Note that although WILD at the SO is not explicitly reported by research made in exceptional LD experiencers indicating a frequency of more than 3 LD per week, PSG-verified REM, substantiated

---

[12] Partial examination of the recorded EEG data for one session, in a preliminary analysis using automated sleep scoring with high accuracy and specificity of detection (F-measure R-stage 0.85 $\pm$ 0.01), performed on a subject attempting OBE at the SO, without known narcolepsy, Periodic Limb Movements Disorders (PLMD) or REM Behaviour disorders (RBD), indicated SOREMP for approximately 9 minutes, 10 minutes after the first cycle of light sleep (N1-2-Stages, not PSG categorised) without hypnagogy or loss of lucidity. Most sessions had a SOREMP prevalence, although some did indicate a direct incidence of REM without N-stages, almost without Sleep Onset Latency (SOL) (a 1-hour meditation was however performed before the attempt). All OBE's attempts had strong hypnic jerks associated with prominent head movements and were accompanied by intracranial sounds.

through Left-Right-Left Right-center (LRLR) eye movement signaling, is typically reported after the first cycle of N1 to N3-stages leading to REM (approx. **90 Minutes**).[13]

It should be emphasized that although the SOREMP states in the above-mentioned unpublished investigation did present neurophysiological characteristics of the R-stage, they did not present the phenomenology ascribed to the powerful imagery perceived in LD, although the subject reported mentation, which could be associated with DLPFC functions. Moreover, the state presents phenomenological similarities with the reported perception of *dream yoga* and *clear light* phenomenon in LD (Johnson, 2017). Mild spontaneous VS-like sensations were also reported in one session although with weak intensity.

The R-stage seen at the SO in such states would otherwise be characteristic of PSG patterns seen in either Narcoleptic patients – which are known to elicit LD, – or is perceived in sleep-deprived population inducing REM rebound at the SO, although excessive daytime sleepiness symptoms generally correlated to narcolepsy were not reported by Tart (1968), or the other subject. As such, it is predicted that the hypothesized SOREMP variation, although presenting similar physiognomies, would not present the pathological characteristic of hypocretin deficiency seen in narcoleptic patients individually measured by PSG Multiple Sleep Latency Test (MSLT).[14] We may further enquire if such state expresses the same absence of normal sympathetic activation during SOREM reported in patients with narcolepsy in such state, when compared to normal R-stages periods.

The theorized correlation of OBE-SO to SOREMP, with a differential diagnosis to Narcoleptic SOREMP, is otherwise partially supported by the recurrent experience of hypnic jerks seen during OBE-onset attempts (Tart, 1967; Crookall, 1980, 1988;

---

[13] Although one study by LaBerge reported a Mnemonic Induced Lucid Dreaming (MILD) 30 minutes after SO.

[14] Criteria by the America Academy of Sleep Medicine (AASM) established the pathological threshold for Narcolepsy type 1 with cataplexy, or hypocretin deficiency syndrome (Cerebrospinal fluid (CSF) measurement of ≤ 110 pg/mL) as a MSLT time of less than 8 minutes with a SOREMP, within 15 minutes of Sleep Onset (SO). In case of OBE subjects, SOREMP can be detected, but without loss lucidity or reported OBE, although frequent hypnic jerks were observed.

Montenegro, 2015), which generally happen during phasic R-stages,[15] although they correspondingly occur in N1-stage as sleep starts. In addition, if hypnic jerks might be associated with excitatory glutamatergic as an equal indication of R-stages, they may be equally indicative of increased gamma activity ($\pm 40\ Hz$). Nevertheless, such conditions need to be better differentiated and dissociated from PLMD, requiring further PSG verification, although research on the subject may be limited, taking into consideration that EMG patterns have poor normative quantification in sleep medicine research (Consens et al., 2005; Schenck, 2005). The association is however still relevant as hypnic jerks are hypothesized to be activated by instability coming from the brainstem reticular formation.

Furthermore, R-stages, assumed as transient OBE states of partial disconnection from the body, would explain how non-ordinary R-states, primarily linked to brain functions, in essence biological, could lead to dreams of telepathy or foretelling, as perceived in LD (Ullman, Krippner and Vaughan, 2002), which, in essence, are rather *psi* phenomena. To the same extent, a clearer understanding of the transient pre-OBE state could provide differential insight into the distinction between LD and OBEs states, analyzing their distinct typology through state differentiation (Montenegro, 2020b).[16]

Non-induced R-stage ISP variations, as a mechanism leading to OBE, would occur during ISP-sleep. Although ISP is known to parallel conditions of R-stage, as predormital hypnagogic-ISP or with SOREMP variations, they are statistically prevalent on average within 1-hour after the Sleep Onset Latency (SOL) or experienced during post-dormital hypnopompic-ISP (SOL < 6 hours) (Girard and Cheyne, 2006). This is a differential diagnostic and conditions which mostly fall outside the SOREMP MSLT pathological definition edge (R-stage ≤ 8 min of SO). Such ISP variation usually presents itself spontaneously during the first R-stage period (after the N-stages) and is mostly associated with a spontaneous increase in arousal during the

---

[15] R-stages are not exclusively tonic phases (atonia) and express transient muscle activity of phasic twitches, although hypnic jerks may not always be OBE related.
[16] Although LD and OBE have a clear differential typology, the condition is often oversighted in current research, but could provide neurophysiological differential insight (Montenegro, 2015).

R-stages offset (e.g., R-stage transitioning to W-stage, upon awakening) (Girard and Cheyne, 2006) and are also known to be triggered by pre-awakening in the N-stages (Takeuchi et al., 1992).

In its pathological variation, RISP is suggested to be mediated by 5-HT hypo-activation (REM-OFF) of the VLPO, although a loss of Orexin 2 (Hypocretin) has been proposed as a counter mechanism to explain it (Lu et al., 2006), with hallucinations suggested to be modulated by 5-HT$_{2A}$ receptors (Jalal, 2018).

The relation is otherwise not new. ISP leading to OBEs (ISP-OBEs) is knowingly prominent in OBE studies, suggesting ISPs are predominant as an OBE induction state or post-OBE, leading to hypnopompic ISP, although the statistical analyzes were provided without PSG examination to ascertain ISP state determination. Even so, statistics of OBE induction associate it with ISP-OBE in **5%** to **72%** of cases (**5%** ($n = 400$) (Green, 1975), **52.57%** ($n = 1115$) (Alegretti and Trivellato, 1999), **72%** ($n = 16185$) (Buhlman, 2001), granting the last two statistics are non-randomized internet surveys and may explain the significant difference). It is equally of note that ISP-VS were reported by less than **41%** ($n = 2397$) (Cheyne, 2002) of ISP experiencers, although little quantification of the phenomena is generally performed in sleep studies, for a further quantification of the relation.

Finally, it is worth noting that such type of experiential analysis has led ISP to be theorized as a mechanism to explain OBEs, although leading theories dismiss OBEs entirely as ISP states. The model is undoubtedly contemptuous of many of the aspects of OBE phenomenology, being inherently reductionist and mostly related to research referring to pathological conditions. The theoretical assumptions are, as such, weakened by trivializing OBE only as pathologies of hallucinatory character, but without PSG experiential analysis, as anecdotally provided here, or more questionably missing explanations for phenomena such as the VS. Either way, the fact suggests that experimentally induced ISP (Takeuchi et al., 1992) do not necessarily convey the transformative power of Near-Death-like Experiences (NDEs) or OBEs, ISP/RISP being mostly perceived as negative,[17] and although a small percentage of ISP-OBE experiences are perceived as positive,

---

[17] The review of such limitations goes beyond the scope of this article.

the condition remains unexplained. Non-pathological ISP-OBEs may as such stem from the referred motor inhibition during increased lucidity in the R-stage offset, with an unspecified neurophysiological predisposition to OBEs (Denis et al., 2015).

SO OBE-C would be associated with W-N-stages transitioning variations. While during W-stages, motoneurons are stimulated by norepinephrine (NA), the voluntary induction of "insensibility" or deep torpidity would hypothetically be prompted by conscious control of tonic immobility possibility mediated by norepinephrine LC neurons (LC-NA) and led by reduced processing of sensory information, while the experiencer would still keep lucidity.

In effect, the state would imply a direct inhibition of the post-synaptic potentials of interneurons of the ventral horn, inhibiting the spinal motor neurons, consequently generating muscle atonia mediated through glutamatergic RSN command. A condition that could be associated with induced or delayed histamine (H_1 and H_2) inhibition at the SO, modulating the TuberoMammillary Nucleus (TMN) which usually promotes excitatory waking effects and consequently preventing a full VLPO GABAergic inhibition and sleep (Figure 5). This is an overall inhibitory condition, otherwise modulated through the downregulation of ARAS brainstem systems, which, in effect, prompts the transitioning between W-state and sleep states, and vice versa.

At least three correlative conditions seem to attest further to the theory. First, the pre-existing state level of relaxation conveyed by psychological assessment of OBE experiencers in many studies provides further evidence of the initial level of the inhibitory state. Granting limitations of the relation, surveys report a predominant relaxing state pre-OBE in **41.3%** to **79%** of correspondents (**41.3%** ($n = 400$), Green, 1975; < **46.92%** ($n = 1115$) (Alegretti and Trivellato, 1999) and **78%** ($n = 339$) (Twemlow, Gabbard and Jones, 1982)). Moreover, the relaxation state is often singularly denoted as being most unusual and out of the ordinary level of relaxation (Green, 1975, p. 50), as reported in *projective* catalepsy.

Second, not only LD is known to occur in N-stages, but awareness is equally experienced at the SO during N1-2-stages of sleep (Foulkes and Vogel, 1965). N-stages lucidity were equally reported in OBE induction studies by Tart (1967, 1968), and N-stage mentation was correspondingly seen in the reported investigation by this author. A condition that could be hypothetically associated with increased Theta-band activity in frontal lobes, known to correlate with the Default Mode Network (DMN) and could equally be prompted by increased proprioception and mind-wandering, as well as reduction of external stimuli perception which might further be associated with increased Theta from the dorsal pathway of the LC.

Third, the hypothetical neurophysiological state of induced catalepsy, dependent on behavioral context, is supported by recent studies indicating that voluntary inhibitory modulation and control of the sympathetic nervous system with increased epinephrine modulation (Kox et al., 2014), and as such the modulation of NA systems in the spinal cord is equally theoretically conceivable. Moreover, NA inhibition of LC neurons are recognized to produce cataleptic states in animal studies. The inferred anecdotal report suggests a state of profound inhibition without phasic twitches with a reported phenomenology associated with sensed presence, among others, in an overall state relatively distinct from the hypothesized OBE-SOREMP induced state, although still without S-VS, a condition ascribed to the relatively little number of attempts.

Finally, we may postulate that the VS could modulate the referred brainstem networks with equivalent inhibitory effects prompting OBEs. Albeit infrequently self-reported, the VS is reported to induce motor paralysis during W-states, with similar symptoms of cataplexy (Montenegro, 2020c). The condition might equally be associated with currently unknown factors associated with the Delta-Gamma PAC seen in W-VS (Figure 1).

Nonetheless, although the hypothesis of transitional states do not provide a specific explanation for the energy-like sensations, granting the correlation is entirely speculative at this point, we may assume that the VS sensations are associated with 5-HT being modulated by endogenous DMT-5-HT$_2$ bidding as further suggested by the OBE-DMT association theory (Strassman, 2001). This is a condition seemingly supported, although indirectly, by the mentioned VS research providing a fascinating outlook at similarities between EEG post-DMT ingestion and the neurophysiological correlates seen in VS-EEG (Montenegro, 2020a), a condition equally correlated with the reported perception of vibrations in post-DMT ingestion.

5 The VS as a transitional state: hypothetical assumptions

The neurophysiological organization of such overlapping and interlocking states provide a framework to hypothesize the VS arising from the LD state or an ISP state, and leading to the reported LD-VS and ISP-VS, including the OBE-C state hypothesis to be mediated by increased Gamma activity (e.g., < 64 Hz), with a similar signature to the one perceived in W-VS (Figure 6).

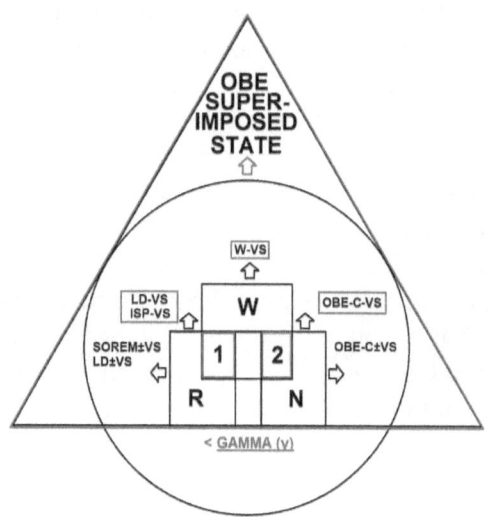

Figure 6 The hypothetical Gamma-wave associated to the VS variations: LD-VS, ISP-VS (W-R-stages variation 1) and OBE-C-VS, hypothesized as N-stage variations and conditions where the onset of the vibrations are seen (SOREMP±VS, LD± and OBE-C±VS) which may lead to the VS and/or the hypothetical superimposed OBE neurophysiological state.

Granting that such hypothetical assumption does not necessarily explain the S-VS occurrences, or indeed, lead to the perception of vibrations and may present different neural signatures, the assumption is nonetheless reasonable considering phenomenological similarities between W-VS, S-VS and VS-OBEs, which are mostly known to differ in experiential intensity, the latter ones reportedly being more intense.

The condition is plausible from a neurophysiological and experiential standpoint. LD states are associated with increased Gamma (40 Hz) (Llinas and Ribary, 1993; Voss et al., 2014) and the report of LD-VS "vibrations" could be, as such, incremented by Gamma waves as they are seen in W-VS above the 40 Hz threshold, leading to OBEs. Furthermore, in most cases, S-VS experiences lead to an increase in arousal and awareness during the experience. Such condition, if further verified, is indeed compatible with a possible increase in Gamma wave frequencies, known to correlate with consciousness and focused arousal (Wang, 2010; Buzsáki and Wang, 2012). Moreover, Gamma waves naturally occur both within R and N-stages, associated with an increase Gamma generation from the pontine nucleus, and although such neural correlates would not equally lead to the same VS typology, they might still induce vibrations-like phenomena – a phenomenon currently under-assessed in sleep research. Finally, Gamma increase is perceived in NDE after experimental cardiac arrest in animal laboratories (Borjigin et al., 2013) and equally reported at 60 Hz during OBE by Tart (1967).[18]

Such a neurophysiological state could hypothetically be followed by an EEG flattening, as recorded in three OBE studies (Tart, 1968; Osis and Mitchell, 1977), consequently being a state evolving from high Gamma to a low voltage Delta state, which might have similar physiological characteristics to N-stages.[19] This is a consciousness-state possibly constituting a transitory, relative cessation of synaptic activity, as reported in patients withdrawing essential care due to endemic illness, hypothesized here as a possible whole mark of a *spirited-away* OBE-state.[20]

---

[18] The 60 Hz Gamma-wave was reported by Tart (1967) in the famous experiment with subject Miss Z., who, in one OBE attempt, correctly perceived a randomly generated 5-digit target number propped in a shelf, above her physical reach.

[19] Although as noted, OBEs may hypothetically arise from OBE-C, SOREMP and ISP-OBE without VS or the hypothetical Gamma increase, nonetheless, the theoretical framework suggests the OBE state is facilitated by Gamma waves.

[20] "Spirited away" is emphasized, as OBE projections may present different characteristics in the so-called *consciousness* projections where, as hypothesized, only

6 Discussion

To support the theoretical assumption of the discussed overlapping and transitional states, well-designed experimental studies are required, taking into account the neurophenomenology of such states, to provide causal neural validity. Altogether, such VS-overlapping states would need to be quantified by excluding the background neural activity associated with each of their own overlapping states (ISP, LD signatures versus ISP-VS and LD-VS and W-VS), to quantify their neural differences and identify their respective neural signatures based on neuroimaging data. If such neurophysiological overlapping states do occur, VS perceptions in each state would need to be assessed through a psychometric phenomenological tool, to allow for better state comparison with W-VS, where a relatively better typology is known. The possibility of this, however, remains still remote considering the current limitations in VS-replication.

Undeniably, if the framework considered here for the OBE-superimposed state is equally a theoretical assumption, it needs to be weighed scientifically against acknowledged OBE theories. In that respect, the proposal of a grounded neuroscientific theory would certainly be need, considering the current 35 theories explaining OBEs (year base 2020) (Montenegro, 2020b). Granting that the aforementioned theories may occasionally provide relevant pathological assessments, like the assumptions proposed here, such theory would need to take into consideration a more extended OBE phenomenology and provide an explanation for them. This author is currently analyzing 85 OBE phenomena, mostly not taken into account in such theories, to be further integrated into the theoretical and experimental scope of OBE's neuroscience state differentiation, based on phenomenology assessment (Montenegro, 2020b).

In addition, even though the theoretical assumption suggested in

---

some elements of the consciousness, such as visual characteristics, are "exterior-ized" without the astral body, although with a different phenomenology than it is seen in remote viewing.

this article steam from naturalistic mechanisms and are speculative, the observations discussed in this article, unlike most of the aforementioned 35 current theories, proposing different theoretical but contradicting OBE models, are not established on pathological disorders or dissociative states. As such, they would at least have the merit, if accurate, to explain OBEs as a natural condition, as they are known to be, thus proposing a non-reductionist, neurophysiologically based theory, although it is acknowledged that the neurophysiological approach remains limited to clarify the referred mechanisms of S-VS, at least without also considering the concurrence of a *vital force*-related mechanism or metaphysical aspects. [21]

Nevertheless, despite the described neurophysiological restrictions, equally-weighted by the lack of information about such states, sleep state research methodology has immensely benefited from contrasting state determination (neurophysiological state determination) with state differentiation (neuropsychological state differentiation) (Mahowald and Schenck, 2005), for states that did not have previously well-defined standard voltage fluctuation differentials, or clear neurophysiology markers. Opposing the neurobiological components (data from neurophysiological research) with the neurophysiological and behavioral states (e.g., psychological, behavioral assessments, psychometric data) was undoubtedly a methodology that provided great insight in early sleep laboratories studies (Foulkes and Vogel, 1965; Foulkes, Spear and Symonds, 1966) and may still prove to help assess the new neurophysiological markers of the VS-states. It is equally undeniable that such methods have led to a better differential diagnostic evaluation of the pathological states of sleep (Crawford et al., 2014).

Finally, the complexity of the neurophysiological OBE state may prove, in itself, to be even more complex to define. Although specific differential neurochemistry may, as stipulated, arise from the research, past OBE research has undoubtedly exemplified the

---

[21] It does not mean either that such an *élan vital*, as termed by French philosopher and Nobel Prize (1927) Henri Bergson, as a vital impetus closely associated with consciousness, would not have a neurochemical base either. A condition which could be associated with 5-HT receptors modulation in the peripheral nervous system and muscle (Montenegro, 2020a).

possible existence of markers that could not be classified into any known sleep stage, even when studied by a leading authority such as Dr. William Charles Dement – an acknowledged pioneer in the field of sleep research.

OBE laboratory studies initially started in 1967, when Charles T. Tart reported complex EEG recordings with distinct neurophysiological EEG patterns of brain activity (Tart, 1967, 1968), which as indicated were partially replicated by this author. Other studies reported similar neurophysiological characteristics but were equally unable to conclusively classify OBE-states as light sleep or REM sleep (Tart, 1968; Hartwell, Janis and Harary, 1974). The neural signatures of such states were referred to as a non-dreaming, "non-awake," non-drowsy patterns (Tart, 1967; Tart, 1968; Hartwell, Joseph and Harary, 1974) and remained unclassified states if compared to the known neurophysiological standards for sleep stages. Furthermore, "alphoid" patterns were predominant in R-stages states (Tart, 1968, p. 10) and N1-stages were described to have poorly developed Theta waves, with unusual lower sleep spindles voltage (patterns typically seen in N2-stages), whereas underdeveloped Theta waves were seen in **64%** of the subjects' sleep pattern spent in borderline states (Tart, 1968). This is a condition indicated to be akin to a state fluctuating between sleep and wakefulness, with well-developed Alpha waves above standard micro voltages norms (Tart, 1968). Even more puzzling conditions were reported describing a lack of Delta waves in N3 stages and stage 4 (old terminology for N3-stage) (Tart, 1968; Janis et al., 1973), all equally controversial sleep neurophysiological markers.

Other polysomnographic conditions had equally contentious results. For example, EOG activity decreased during OBEs, and although such changes were statistically significant (Tart, 1968; Janis et al., 1973; Osis and Mitchell, 1977; Gabbard and Twemlow, 1984), they were not reported in other studies (Hartwell, 1973). EMG fluctuations reported by psychiatrists Gabbard and Twemlow (1984), indicating synchronized EEG patterns, had no tonic-phasic variability during OBEs, contrary to the regular tonic and phasic autonomic functions, also reported by Hartwell, Janis, and Harary (1974).

If a lack of technical specification impairs the analysis of such findings, they grant the need for further systematic research of such states, with gifted OBE subjects. They certainly would require further replication to provide state-of-the-art contextual data analysis with a state differentiation context. Nonetheless, even if such findings were never replicated and remain with elusive descriptions, they nevertheless provide a preview of the complexities ahead, in the understanding of neurophysiological state differentiation in non-ordinary sleep states research. We believe that such complexities need to be integrated into the actual research frame of consciousness research, specifically if it is to be considered as a modern scientific research field capable of further developing in the future.

5 Conclusion

Fundamentally, future research will need to provide further neurophysiological determinants of a neural VS signature across the sleep stages, to reach the criteria of evidence needed to achieve state differentiation. The notion of state differentiation presented here is considered, as such, vital for the formulation of a theory that would scientifically validate the VS as novel state, possibly bringing OBE neuroscience at the forefront of consciousness research — bridging the existing gap in the current research divide between a purely phenomenological approach and a more neurophysiological one.

Certainly, it is hoped that future experiential analysis providing hard data of the neurophysiological correlates of the VS associated with OBE phenomena will stimulate the engagement of the scientific community in studying them in a new light, promoting neurophenomenology as a non-reductionist and essential scientific approach at the forefront consciousness research.

The research in question may equally provide insightful knowledge and lead to a more essential understanding of how to induce such non-ordinary states, through the design of scientifically based techniques taking into consideration neuroscience knowledge. This is a research field that may further provide useful data for the development of EEG-feedback headband (high quality consumer-grade products), which going forward, may stimulate the induction of the VS by providing neophyte VS-experiencers a way to objectify their subjective experience and measure the characteristics of their own subjective states, in full autonomy. Precisely so, if such headbands can characterize their experiences through a comparative analysis with the results obtained by more experienced VS practitioners.

Such research may also shift the current nosological framework associated with the OBE field. OBEs, ISP-OBE and VS are often described as hallucinatory states. However, research providing support to the notion of Gamma lucidity ascribed to the accompanying states may convey more convincing evidence to suggest that VS-OBEs and VS-ISP, leading to OBEs that are known to be experienced within a highly engaging and vivid frame of mind, are rather mind-expanding than hallucinatory experiences, contrary to current appraisal.

Similarly, if the transitional states to OBE are not assumed to be their unique neurophysiological triggers, the current theoretical framework nonetheless indicates that *Vibrational Meditation* training is essential to elicit OBEs with a high expression of awareness. Such type of mental training may increase neural potentiation leading to a neuro-plastic adaptation, consequently eliciting a higher frequency of spontaneous transformative states of being (Hebb's rule), as undeniably suggested by experiential research analysis (Vieira, 1986; Montenegro, 2015; Trivellato, 2015).

More importantly, such training could consequently help those emotionally affected by negative ISP experiences to transform them into more blissful, uplifting, and enlightening doors of perception to the OBE realms, as they are reported to be. The same condition could apply to help trigger OBEs starting from R-stages, equally providing a higher experiential understanding of the phenomenological differences between LD and OBEs, which are commonly mistaken as being the same experience.

Assuredly, future research will "challenge" the foundation of consciousness theories such as the microtubule proposed by Hameroff (Craddock et al., 2012), allowing to test, for example, neural microtubules post-translational stability and their electrical oscillation during VS or OBE, which in turn may foster understanding into brain-consciousness mechanisms. Future VS research might also be empirically tested through Diffusion Tensor

Imaging (TDI), allowing for tracking subcortical relays associated with VS-induction and through the use of Functional Connectivity MRI (fcMRI), mapping the networks changes associated to VS-OBEs. Again, this may provide insight into the nature of brain changes associated with the VS and their evolutionary possibilities.

Last but not least. A research focus on the VS may be better encouraged by research in the field of neurodegenerative diseases and the prevention of early dementia and Alzheimer, where such states could provide exciting applications. Undeniably, beside the broad role of Gamma wave in cognition and insight (Buzsáki and Wang, 2012), amongst others, decreased Gamma synchrony and reduced Gamma power have been known to be present in mild cognitive impairments and is existent in dementia onset and Alzheimer Disease (Herrmann and Demiralp, 2005; Koenig et al., 2005; Gillespie et al., 2016). The research into the VS could indeed provide immense benefit in preventing, if not reverting symptoms, of such pathologies, thus achieving a higher universal reach in its scope, more specifically so if EEG-neurofeedback training tools are provided to such necessitous population.

Acknowledgments

The author would like to express his profound gratitude to Dr. Peter Fenwick M.D., F.R.C. Psych., fellowship of the Royal College of Psychiatrists, for his mentoring and the support provided allowing this author to carry consciousness research to its next level.

Extracts from that essay were compiled from the dissertation the author wrote for the School of Advanced Education and Research (SAERA) affiliated to Isabella I University.

Bibliography

Abraham, T. (2010). The Emerging Field of Neuroconscientiology: the Challenges Facing and Opportunities for Exploring a New Frontier of Conscientiological Research. Journal of Conscientiology, July-December, Vol.

13(49), pp. 41-54.

Alegretti, W. (2008). A Research Approach to the Vibrational State through the Study of Brain Activity. Journal of Consciousness, October-December, Vol. 11(42), pp. 221-255.

Alegretti, W. and Trivellato, N. (1999). Survey Research about the Projection of the Consciousness through the internet. (Eds) 1st International Forum of Consciousness Research and 2nd International Congress of Projectiology. Rio de Janeiro: IIPC Ed.

Almansori, M., Ijaz, M. and Ahmed, S. (2006). Cerebral arrhythmia influencing cardiac rhythm: A case of ictal bradycardia. Seizure, 15(6), pp. 459-461.

Aru, J., Suzuki, M., Rutiku, R., Larkum, M. and Bachmann, T. (2019). Coupling the State and Contents of Consciousness. Frontiers in Systems Neuroscience, 13(Article 43), pp. 1-9.

Bartlam, R. and Mohanraj, R. (2016). Ictal bradyarrhythmias and asystole requiring pacemaker implantation: Combined EEG–ECG analysis of 5 cases. Epilepsy & Behavior, 64, pp. 212-215.

Borjigin, J., Lee, U., Liu, T., Pal, D., Huff, S., Klarr, D., Sloboda, J., Hernandez, J., Wang, M. and Mashour, G. (2013). Surge of neurophysiological coherence and connectivity in the dying brain. Proceedings of the National Academy of Sciences, 110(35), pp. 14432-14437.

Brownstone, R. and Chopek, J. (2018). Reticulospinal Systems for Tuning Motor Commands. Frontiers in Neural Circuits, 12(Article 30), p. 10.

Buhlman, W. (2001). The Secret of the Soul: using Out-of-Body Experiences to understand our true nature. 1st ed. New York: Harper Collins, p. 271.

Buzsáki, G. and Wang, X. (2012). Mechanisms of Gamma Oscillations. Annual Review of Neuroscience, 35(1), pp. 203-225.

Callaway, J., McKenna, D., Grob, C., Brito, G., Raymon, L., Poland, R., Andrade, E. and Mash, D. (1999). Pharmacokinetics of Hoasca alkaloids in healthy humans. Journal of Ethnopharmacology, 65(3), pp. 243-256.

Chawla, L. and Seneff, M. (2013). End-of-life electrical surges. Proceedings of the National Academy of Sciences, 110(44), pp. E4123-E4123.

Cheyne, J. (2002). Waterloo Unusual Sleep Experiences Questionnaire – VIIIa Technical Report. [ebook] University of Waterloo. Available at: http://citeseerx.ist.psu.edu/viewdoc/summary?doi=10.1.1.394.7765.

Cheyne, J. (2010). Recurrent isolated sleep paralysis. In: Thorpy, M. and Plazzi, G. (2010). The parasomnias and other sleep-related movement disorders. 1st ed. Cambridge, UK: Cambridge University Press, p. 341.

Coffey, M. (2008). Explorers of the infinite: the secret spiritual lives of extreme athletes - and what they reveal about near-death experiences, psychics communication, and touching the beyond. 1st ed. London: Penguin Group, p. 288.

Consens, F., Chervin, R., Koeppe, R., Little, R., Liu, S., Junck, L., Angell, K., Heumann, M. and Gilman, S. (2005). Validation of a Polysomnographic Score for REM Sleep Behavior Disorder. Sleep, 28(8), pp. 993-997.

Cornillier, P. (1927). La Survivance de l'âme et son évolution après la mort - Comptes rendus d'expériences. 1st ed. Paris: Librarie Félix Alcan, p. 646.

Craddock, T., St. George, M., Freedman, H., Barakat, K., Damaraju, S., Hameroff, S. and Tuszynski, J. (2012). Computational Predictions of Volatile

Anesthetic Interactions with the Microtubule Cytoskeleton: Implications for Side Effects of General Anesthesia. PLoS ONE, 7(6), p. e37251.

Crawford, M., Espie, C., Bartlett, D. and Grunstein, R. (2014). Integrating psychology and medicine in CPAP adherence – New concepts?. Sleep Medicine Reviews, 18(2), pp. 123-139.

Crookall, R. (1978). What Happens When You Die. 1st ed. London: London Colin Smythe, p. 208.

Crookall, R. (1980). Case-book of astral projection. 2nd ed. Secaucus, N.J.: University Books, p. 160.

Crookall, R. (1988). The study and practice of astral projection. 2nd ed. Secaucus, N.J.: Citadel Press, p. 234.

De Koninck, J. (2012). Sleep, Dreams, and Dreaming. Eds: Morin, C. and Espie, C. (2012). The Oxford handbook of sleep and sleep disorders. 1st ed. Oxford: Oxford University Press, p. 871.

De Rochas D'Anglun, A. (1895). L'extériorisation de la sensibilité. 2nd ed. Paris: Chamuel Éditeur, p. 284.

Denis, D., French, C., Rowe, R., Zavos, H., Nolan, P., Parsons, M. and Gregory, A. (2015). A twin and molecular genetics study of sleep paralysis and associated factors. Journal of Sleep Research, 24(4), pp. 438-446.

Descartes, R. (1894). Le Discours de la méthode - bien conduire sa raison, et chercher la vérité dans les sciences. Paris: Libraire de la Bibliothèque Nationale, p. 135.

Durville, H. (1909). Le Fantôme des Vivants: Anatomie et Physiologie de l'Âme: Recherches Expérimentales. 1st ed. Paris: Librairie du Magnétisme, p. 360.

Evans-Wentz, W. (1958). Tibetan yoga and secret doctrines. 1st ed. Oxford: Oxford University Press, p. 432.

Everett, H. (1963). Sleep Paralysis in medical students. The Journal of Nervous and Mental Disease, 136(3), pp. 283-287.

Fitzgerald, M., Gruener, G. and Mtui, E. (2012). Clinical neuroanatomy and neuroscience. 6th ed. [Edinburgh]: Saunders Elsevier, p. 391.

Foulkes, D. and Vogel, G. (1965). Mental activity at sleep onset. Journal of Abnormal Psychology, 70(4), pp. 231-243.

Foulkes, D., Spear, P. and Symonds, J. (1966). Individual differences in mental activity at sleep onset. Journal of Abnormal Psychology, 71(4), pp. 280-286.

Fox, O. (1962). Astral projection – A record of Out-of-Body Experiences. Don Mills, Ontario: Citadel Press, p. 160.

Freud, S. (2010). The interpretion of dreams. New York: Basic Books, p. 677.

Gabbard, G. and Twemlow, S. (1984). With the eyes of the mind - An empirical analysis of Out-of-Body States. 1st ed. New York: Praeger Publisher, p. 272.

Gabbard, G., Twemlow, S. and Jones, F. (1982). Differential Diagnosis of Altered Mind/Body Perception. Psychiatry, 45(4), pp. 361-369.

Gillespie, A., Jones, E., Lin, Y., Karlsson, M., Kay, K., Yoon, S., Tong, L., Nova, P., Carr, J., Frank, L. and Huang, Y. (2016). Apolipoprotein E4 Causes Age-Dependent Disruption of Slow Gamma Oscillations during Hippocampal Sharp-Wave Ripples. Neuron, 90(4), pp. 740-751.

Girard, T. and Cheyne, J. (2006). Timing of spontaneous sleep-paralysis episodes. Journal of Sleep Research, 15(2), pp. 222-229.

Green, C. (1975). Out-of-Body Experiences. 2nd ed. New York: Ballantine

Books, p. 170.

Hackforth, R. (1972). Plato's Phaedo. 1st ed. Cambridge: Cambridge University Press, p. 207.

Hamill, J., Hallak, J., Dursun, S. and Baker, G. (2019). Ayahuasca: Psychological and Physiologic Effects, Pharmacology and Potential Uses in Addiction and Mental Illness. Current Neuropharmacology,, 17(2), pp. 108-128.

Hamilton, W. (1973). Phaedrus & The Seventh And Eighth Letters. Cambridge: Penguin Classics, p. 160.

Hartwell, J., Janis, J. and Harary, B. (1974). A study of the physiological variables associated with out-of-body experiences. (Eds.) J. D. Morris, W. G. Roll, & R. L. Morris; Research in Parapsychology: The Scarecrow Press, pp. 127-129.

Hermann, C. and Demiralp, T. (2005). Human EEG gamma oscillations in neuropsychiatric disorders. Clinical Neurophysiology, 116(12), pp. 2719-2733.

Hervey de Saint-Denys, L. (2005). Les rêves et les moyens de les diriger. Genève: Éd. Arbre d'Or, p. 264.

Hobson, J., Lydic, R. and Baghdoyan, H. (1986). Evolving concepts of sleep cycle generation: From brain centers to neuronal populations. Behavioral and Brain Sciences, 9(3), pp. 371-400.

Janis, J., Hartwell, J., Harary, B., Levin, J. and Morris, R. (1973). A description of the physiological variables connected with an out-of-the body study. 1st ed. (eds.) J. D. Morris, W. G. Roll, & R. L. Morris; Research in Parapsychology: The Scarecrow Press, pp. 36-37.

Johnson, C. (2017). Llewellyn's Complete Book of Lucid Dreaming. Kindle Edition: Llewellyn Worldwide, p. 379.

Kinnier Wilson, S. (1928). The Narcolepsies. Brain, 51(1), pp. 63-109.

Koenig, T., Prichep, L., Dierks, T., Hubl, D., Wahlund, L., John, E. and Jelic, V. (2005). Decreased EEG synchronization in Alzheimer's disease and mild cognitive impairment. Neurobiology of Aging, 26(2), pp. 165-171.

Kompanje, E. (2008). 'The devil lay upon her and held her down'Hypnagogic hallucinations and sleep paralysis described by the Dutch physician Isbrand van Diemerbroeck (1609-1674) in 1664. Journal of Sleep Research, 17(4), pp. 464-467.

Kox, M., van Eijk, L., Zwaag, J., van den Wildenberg, J., Sweep, F., van der Hoeven, J. and Pickkers, P. (2014). Voluntary activation of the sympathetic nervous system and attenuation of the innate immune response in humans. Proceedings of the National Academy of Sciences, 111(20), pp. 7379-7384.

Llinas, R. and Ribary, U. (1993). Coherent 40-Hz oscillation characterizes dream state in humans. Proceedings of the National Academy of Sciences, 90(5), pp.2078-2081.

Lu, J., Sherman, D., Devor, M. and Saper, C. (2006). A putative flip–flop switch for control of REM sleep. Nature, 441(7093), pp. 589-594.

Mahowald, M. and Schenck, C. (1999). Dissociated States of Wakefulness and Sleep. (Eds.) Lydic, R. and Baghdoyan, H. (1999). Handbook of behavioral state control: cellular and molecular mechanisms. 1st ed. Boca Raton: CRC Press, p. 699.

Mahowald, M. and Schenck, C. (2005). Insights from studying human sleep disorders. Nature, 437(7063), pp. 1279-1285.

Manni, R. (2005). Rapid eye movement sleep, non-rapid eye movement sleep, dreams, and hallucinations. Current Psychiatry Reports, 7(3), pp. 196-200.

Maury, L. (2013). Le sommeil et les rêves Format Kindle de Alfred Maury. Kindle Edition: Ebox Editions.

Monroe, R. (1971). Journeys out of the Body. 1st ed. New York: Doubleday Editions, p. 280.

Montenegro, R. (2015). The out-of-body Experiences – An Experiential Anthology. 1st ed. Livros e Puplicações, p. 339.

Montenegro, R. (2020a). The neurophysiological correlates of the Vibrational State. Manuscript in preparation.

Montenegro, R. (2020b). State Differentiation and Neurophysiological Correlates of Out-of-the Body Experiences: towards a comprehensive theory. Manuscript in preparation.

Montenegro, R. (2020c). A brainstem OBE theory. Manuscript in preparation.

Morin, C. and Espie, C. (2012). The Oxford handbook of sleep and sleep disorders. 1st ed. Oxford: Oxford University Press, p. 871.

Muldoon, S. and Carrington, H. (1929). Projection of the Astral Body. 1st ed. London: London Rider & Co., p. 242.

Osis, K. and Mitchell, J. (1977). Physiological correlates of reported out-of-the body Experiences. Journal of the Society for Psychical Research, 49(772), pp. 525-536.

Recreational Use, Analysis and Toxicity of Tryptamines. (2015). Current Neuropharmacology, 13(1), pp. 26-46.

Rimpoche, S. (2012). The Tibetan Book Of Living And Dying. [Kindle] Peguin Random house.

Rinella, M. (2000). Supplementing the Ecstatic: Plato, the Eleusinian Mysteries and the Phaedrus. Polis: The Journal for Ancient Greek Political Thought, 17(1-2), pp. 61-78.

Rinpoche, S. (2012). The Tibetan Book Of Living And Dying. [Kindle] Penguin Random house.

Rodrigues Pinheiro, R. (2013). Correlatos Eletroencefalográficos do Estado Vibracional. Master of Psychobiology. Universidade Federal do Rio Grande do Norte.

Rogo, D. (1986). Leaving the body – A complete guide to astral projection. 2nd ed. New York: Prentice Hall Press, p. 190.

Salley, R. (1982). REM Sleep Phenomena During Out-of-Body Experiences. The Journal of the American Society for Psychical Research, 76, pp. 157-165.

Schenck, C. (2005). Clinical and Research Implications of a Validated Polysomnographic Scoring Method for REM Sleep Behavior Disorder. Sleep, 28(8), pp. 917-919.

Schomer, D. and Lopes da Silva, F. (2011). Niedermeyer's electroencephalography. 6th ed. Philadelphia: Wolters Kluwer Health/Lippincott Williams & Wilkins, p. 1306.

Solomonova, E. (2018). leep Paralysis: Phenomenology, Neurophysiology, and Treatment. in E. (2018). S in Christoff, K. and Fox, K. (Eds). The Oxford handbook of spontaneous thought - mind-wandering, creativity, and dreaming: Oxford University Press, pp. 435-456.

Stern, J. and Engel, J. (2013). Atlas of EEG patterns. 2nd ed. Philadelphia: Wolters Kluwer/Lippincott Williams & Wilkins Health, p. 457.

Strassman, R. (2001). DMT: the spirit molecule: a doctor's revolutionary research into the biology of near-death and mystical experiences. 1st ed. Rochester: VT: Park Street Press, p. 320.

Swedenborg, E. (1884). Heaven and Its Wonders and Hell From Things Heard and Seen. West Chester, Pennsylvania: Swedenborg Foudnation, p.574.

Swedenborg, E. (1918). Emanuel Swedenborg's Journal of dreams and spiritual experiences in the year seventeen hundred and forty-four. Bryn Athyn, Pa, p. 108.

Takakusaki, K., Chiba, R., Nozu, T. and Okumura, T. (2015). Brainstem control of locomotion and muscle tone with special reference to the role of the mesopontine tegmentum and medullary reticulospinal systems. Journal of Neural Transmission, 123(7), pp. 695-729.

Takeuchi, T., Miyasita, A., Sasaki, Y., Inugami, M. and Fukuda, K. (1992). Isolated Sleep Paralysis Elicited by Sleep Interruption. Sleep, 15(3), pp.217-225.

Tart, C. (1967). A second Psychophysiological study of out-of-the body experiences in a gifted subject. Parapsychology, 9, pp. 251-258.

Tart, C. (1968). A Psychophysiological Study of Out-of-the-Body Experiences in a Selected Subject. Journal of the American Society for Psychical Research, 62(1), pp. 3-27.

Terzano, M., Parrino, L. and Spaggiari, M. (1988). The cyclic alternating pattern sequences in the dynamic organization of sleep. Electroencephalography and Clinical Neurophysiology, 69(5), pp. 437-447.

Theochari, A., P. (2008). Artemidorus's Oneirocritica. Dream Analysis in the 2nd Century A.D. Hellenic Psychiatric Hospital, 5(2), pp. 83-85.

Trivelatto, N. (2014). Vibrational State: Qualitative and Quantitative Analysis. Journal of Consciousness, 18(59), p. 337.

Trivellato, N. (2008). Measurable Attributes of the Vibrational State Technique. Journal of Conscienciousness, 11(42), p. 165.

Trivellato, N. (2015). Estado Vibracional - Pesquisas, técnicas e aplicações. 1st ed. Évoramonte: International Academy of Consciousness, p. 507.

Ullman, M., Krippner, S. and Vaughan, A. (2002). Dream Telepathy - Experiments in nocturnal sensory perception. 1st ed. Charlottesville: Hampton Road Publishing House, p. 274.

Verene, D. (2016). Metaphysics and the Modern World. Cascade Books, p.143.

Vieira, W. (1981). Projeções da Consciência - Diário de Experiências Fora do Corpo Físico. 1st ed. Edi. IIP, p. 268.

Vieira, W. (1986). Projectiologia - Panorama das experiências fora do corpo. 1st ed. Rio de Janeiro: Edição do Autor, p. 928.

Vieira, W. and Xavier, F. (1958). Evolução em dois mundos. 1st ed. Rio de Janeiro: Federação Espírita Brasileira, p. 219.

Voss, U., Holzmann, R., Hobson, A., Paulus, W., Koppehele-Gossel, J., Klimke, A. and Nitsche, M. (2014). Induction of self awareness in dreams through frontal low current stimulation of gamma activity. Nature Neuroscience, 17(6), pp. 810-812.

Waggoner, R. (2009). Lucid dreaming. 2nd ed. Needham, Mass.: Moment Point Press, p. 320.

Wang, X. (2010). Neurophysiological and Computational Principles of Cortical Rhythms in Cognition. Physiological Reviews, 90(3), pp. 1195-1268.

Wasson, R., Hofmann, A. and Ruck, C. (2008). The Road to Eleusis: Unveiling the Secret of the Mysteries. 13th ed. North Atlantic Books, p. 151.

Wulff, K., Gatti, S., Wettstein, J. and Foster, R. (2010). Sleep and circadian rhythm disruption in psychiatric and neurodegenerative disease. Nature Reviews Neuroscience, 11(8), pp. 589-599.

AutoRicerca

Subtle energies or subtle matters?

Massimiliano Sassoli de Bianchi

Issue 20
Year 2020
Pages 233-265

Abstract

The concept of energy is central in all of modern science and is obviously of great importance also in the study of the psycho-energetic phenomena. However, there still is some confusion about a proper understanding of this concept. The purpose of the present article, mainly educational, is to provide a correct interpretation of the concept of energy and of its transport, in the different physical systems, to facilitate the formulation of scientifically well-posed questions, especially in the study of the energetic dynamics associated with the controversial "subtle paramatters."

AutoRicerca, Issue 20, Year 2020, Pages 233-265

1 Introduction

Energy is a fundamental concept, not only for the mainstream re-searchers, who calculate and measure its exchanges between the different physical systems in laboratories of chemistry, physics and biology, but also for the so-called self-researchers, investigating the psycho-energetic phenomena, i.e., the mysterious "subtle" "forms of energy,"[1] also called, depending on the context, bioenergies, consciential energies, extraphysical energies, prana, chi, orgone, etc. (Tiller, 1993; Zamperini, 1998; Abs de Lima, 2005; Bruce, 2007; Sassoli de Bianchi, 2009a).

These energies would be at the origin of controversial phenomena such as *psychokinesis* (PK), *spiritual healing, remote viewing,* and more generally of the multiple manifestations of the consciousness, beyond the limits of our physical body (Vieira, 2002).

The writer has a foot in both camps, being both a theoretical physicist and a self-researcher, also dedicated to a better understanding and self-experimenting with these non-ordinary "forms" of energy, whose reality remains entirely hypothetical and still extremely controversial within the academic world. Based on this twofold perspective, I can say without hesitation that there are many confusions, both on the part of conventional researchers, about the nature of these "subtle energies," and on the part of the non-conventional self-researchers, about a proper understanding of the basic concept of *energy*, and its possible application in the description of the parapsychic phenomena governed by human intentionality.

The purpose of this article is to provide a sort of conceptual map, of an elementary level, on the important issue of energy, so that those who today are interested (as students, teachers and/or researchers) in "subtle energies," will be able to operate those important distinctions without which it will be difficult to clarify, both theoretically and experimentally, such a vast and delicate subject.

The term "subtle energy," as it will become clear from reading this

---

[1] As you will discover from reading this article, the concept of "form of energy" is misleading. This is why the term is here in quotes.

article, is inappropriate. This is so not so much because the adjective "subtle" might be misleading in some cases, its meaning being only metaphorical, but mainly because:

It makes no sense to qualify energy, since there is one and only one form of energy, and not different forms of energy.

It should be noted that many investigators working in the field of inner research, who are involved in the study of the psycho-energetic phenomena, do not necessarily possess a specific preparation in the field of physics. So, a certain level of confusion stems from an insufficient understanding of this fundamental branch of knowledge. This will produce some basic confusions, such as mixing up, for example, the concept of *force* with that of *energy*. Some people will speak, in an erroneously interchangeable way, of "vital force" and "vital energy." Strictly speaking, however, if "force" and "energy" are different *physical quantities*, it is certainly desirable also to distinguish the related concepts of "vital force" and "vital energy," and possibly explain in what they would differ, so much as to deserve different names.

On the other hand, I have personally been able to ascertain that a lot of confusion is sometimes conveyed also by researchers having a more solid scientific background, if not by physicists. In this case the confusion is obviously more subtle, as no longer attributable to a lack of specific knowledge about the subject, but rather to an insufficient reflection about its conceptual foundations. Unfortunately, in physics, as in the evolution of biological systems, some real "living fossils" exist, that despite their age mysteriously continue to replicate (Hermann & Job, 1996). These fossils may become formidable obstacles, especially when certain basic concepts must be applied to new fields of investigation, whose phenomenology is still unstable and difficult to demarcate, as is the case of psycho-energetics, and this the more so when many of the researchers working in these fringe areas have a modest scientific background.

I thus believe that the conceptual clarification proposed in this work, despite being truly elementary, can be quite advantageous, not only to those who are totally ignorant of physics, but also to those who, despite having a more solid scientific culture, or a very solid one, have never reflected deeply, or deeply enough, about the meaning of certain basic notions, like the one of *energy* and of its exchange mechanisms.

In the discussion that follows, I will almost totally avoid the use of mathematical formulas, in order not to discourage those readers who still maintain to this day, unfortunately, a strong idiosyncrasy toward formal languages, even though these are obviously needed to express with the necessary precision certain concepts and their relations. On the other hand, for completeness, I will present a simple mathematical relation in the Appendix.

The article is structured as follows: I will start by presenting what are the basic concepts that it is important to know, and to distinguish, in relation to the theme of energy and, more particularly, in relation to its flow between different physical systems. I will try above all to clarify the difference between *material substances* and *immaterial substances*, between *energy* and *energy carriers*, highlighting some of the most pernicious confusions. To facilitate the understanding, I will make use of many elementary examples.

Thanks to this conceptual clarification, I will then address a few well-posed questions, in relation to the theme of psycho-energetics, i.e., of the exchanges of so-called "subtle energies." I will explain also why many historical terms of physics, and consequently many self-research neologisms (like those proposed in Conscientiology[2]) are inappropriate, in the sense of being potentially misleading, and therefore should be avoided wherever possible (and replaced with more appropriate terms). I will also spend a few words on the generalization of the concepts presented in the case where the behavior of the material substances under consideration is not classical (in a sense that I will make precise), but for example quantum, or quantum-like.

2 Material and immaterial substances

Let me start by defining some concepts. By the term of *material substance*, or simply of *matter* (not to be confused, as we shall see, with

---

[2] According to its proponent *Waldo Vieira* (2002), Conscientiology is meant to be the science studying consciousness in an integral, holosomatic, multidimensional, multimillenary, multiexistential manner and, above all, according to its reactions with regard to immanent energy, consciential energy and its own multiple states.

the concept of *mass*), I will refer in this article to the *substratum* of *physical entities*, i.e., to "the stuff physical entities are made of." In order not to complicate too much the discussion at a conceptual level, in the following I will only consider material substances of a *classical* nature, i.e., matters having the special property of *being present at all times in our ordinary three-dimensional physical space* (I will say more about non-classical substances, like quantum substances, later on).

A material substance must therefore be understood as an entity to which one can attribute certain *properties*, namely *physical properties*. Some of these properties will characterize the very identity of the substance, while others will determine its *state*, that is, its specific *condition*, at a given moment.

One of the main characteristics of classical physical entities is, as I just pointed out, to be always present, that is "contained," in our *ordinary physical space* (for simplicity, I will hereafter simply use the term "space," meaning by it the ordinary three-dimensional space, which is only a small portion of the totality of our physical space). This means that material substances can be contained in certain regions of space, and that it makes sense to speak of the *quantity of a given material substance* (or quantity of matter) present in a given region, as it also makes sense to speak of the flow of a material substance that enters and exits a given region of space, or the flow of a specific substance that is transferred from a physical entity to another physical entity.

Conceptually speaking, it is important to make a clear *ontological distinction* between two different categories: the category of *material substances* and the category of *immaterial substances* (or theoretical substances, abstract substances, etc.). This distinction between "material substances" and "immaterial substances" has to do with the distinction between "material substances" and the "properties of material substances."

Let me explain. We can generally say that a substance has or has not a particular property. For example, the material substance "wood" has the property of "being burnable," but does not possess the property of "being a good electrical conductor." There are however particular classes of properties that certain substances may possess not only *qualitatively* (in the sense of having or not having them), but also *quantitatively*, in the sense that they can possess a certain *quantity* of them, which may vary depending on circumstances. In other words, these are properties that can be described in terms of *content*, and therefore behave *as if* they were material substances, although in fact they are not, being instead *properties of material substances*. One could say that they are *substance-like* properties, since they behave similarly to material substances, even if they aren't such.

Energy is perhaps the most typical, and certainly one of the most important, examples of an immaterial substance. Material substances do in fact possess energy (material substances without energy are not known) and they can possess a variable amount of it, i.e., they are able to *contain* a more or less considerable amount of the substance-like property "energy," depending on their state and context. Also, akin to a material substance, energy can *flow* (move, be transferred, etc.) from one spatial region to another, and more generally from one physical entity to another physical entity. The same holds true for many other properties besides energy, which in physics are usually called *physical quantities*, such as *momentum* (either translational or angular), *electric charge* and *entropy*, to name only the better-known ones.

Some immaterial substances, such as energy, linear momentum, angular momentum and electric charge, are *conserved* quantities. This means that they can neither be created nor be destroyed, but only transferred from one entity to another. And of course, they can also be stored within the different physical entities. In other words, in the same way one can speak of the flow of a material substance, like *water*, for example from one container to another container, so it is possible to speak of the flow of energy from one system to another, or of the flow of momentum, electric charge, entropy, etc. However, these flows are associated – I repeat it once again, since it is an important point – to *immaterial substances*, whose behavior certainly resembles that of material substances, but this doesn't mean they have to be considered as such. It is indeed about a "flow of *properties* of material substances," and not a "flow of material substances."

Energy being a property of material substances, it is a sort of supervening aspect of our reality. It exists, so to speak, just because there is a universe of material substances that can carry it. Exactly as for the Italian language, which only exists because there are a number of different material supports allowing for its manifestation. But apparently it cannot exist autonomously, i.e., regardless of these material supports. In other words, the existence of substance-like immaterial entities such as energy appears to be bound to the existence of the material substances that support them.

Not all immaterial substances are however conserved. *Entropy* for example, can be created from nothing in a physical system, though it can never be destroyed (unless evidence to the contrary). On the other hand, material substances, depending on circumstances, can be conserved, created or destroyed. A typical example is that of *chemical* or *nuclear reactions*, during which certain material substances are transformed into others, hence there is a double process of creation and destruction. In other words, the *quantity of a specific material substance* is in general not conserved and may therefore either increase or decrease in the course of a specific process.

3 Energy and energy carriers

In this paper, I will mainly focus on the immaterial substance called "energy," which as is well known is always conserved in physical processes, in the sense that the amount of energy contained in a given region of space may change if and only if a *current of energy* flows across the surface of that region. Similarly, the amount of energy contained in a physical entity can increase (decrease) if and only if such an entity absorbs (emits) energy, in an exchange with its external environment. The *intensity of the energy current*, usually symbolized by the capital letter P, corresponds to what is conventionally referred to as the *power*. In general, the intensity of the current of a given substance (whether material or immaterial), equals the *quantity of substance that flows through a given area per unit time*. When the current is zero, this simply means that the substance remains stationary (with respect to a given referential), i.e., that it doesn't flow.

But let me consider now what are the modalities with which energy can flow, in general terms, from a physical entity to another physical entity. For this, it is necessary to distinguish 5 basic concepts (see the energy flow diagram of Figure 1):

1. The physical entity which is the *source* of energy (S);
    The physical entity which is the *receiver* of energy (R);
2. The *material* substance (M) which is the *carrier* of energy;
3. The *immaterial* substance (I) which is the *carrier* of energy;
4. The immaterial substance *energy* (E).

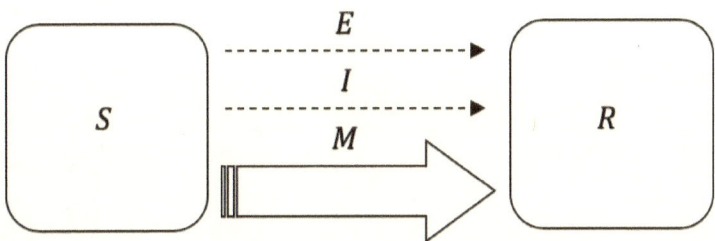

Figure 1 An *energy flow diagram* schematically describing a process of transfer of energy E from a source S to a receiver R, using a material carrier M (indicated by a solid arrow) and an immaterial carrier I (indicated by a dashed arrow).

It is important to note that in a process of energy transfer between a source S and a receiver R, a material carrier M is necessarily always present. The following cases can however be distinguished:

A. The material substance M flows from S to R, and is the only substance carrying energy;
B. The material substance M flows from S to R, but is not the only substance carrying energy, which is also carried by one or more immaterial substances;
C. The material substance M does not flow from S to R (the current is zero) and energy is carried only by one or more immaterial substances.

To understand the reason for the distinction of these 3 cases, and especially the distinction between the immaterial substance "energy" and its carriers, which can be either material or immaterial substances, the best way to proceed is to consider some concrete examples, to illustrate the different mechanisms involved.

4 A few illustrative examples

(1) **Hand-ball-pins.** *S* is a hand, *R* consists of the pins placed on a bowling alley, *M* is a bowling ball and *I* is momentum.

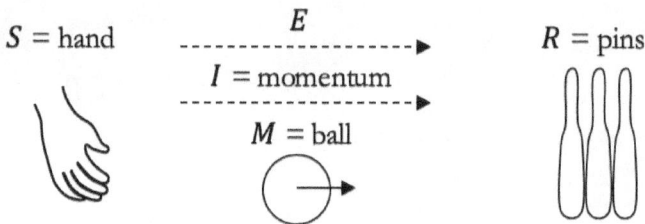

Figure 2 The *energy flow diagram* schematically describing a process where a bowling player transfers energy from his/her hand to the pins, using for this the material carrier "ball" and the immaterial carrier "momentum."

More precisely, *S* communicates to the ball a certain *momentum*, and since a body in motion carries energy, by doing so it transfers to the ball a certain amount of energy. In other words, between *S* and *R* both a material substance (the stuff the ball is made of) and an immaterial substance (the momentum carried by the ball) flow. When the ball comes into contact with the pins, it transfers to them part of the momentum it carries, and in this way also part of its energy (putting them in motion).

The ball is therefore the material carrier of momentum, and momentum is the immaterial carrier of energy.

In this example, besides the presence of a current of energy and of momentum (two immaterial substances), there is also the presence of a current of matter: the substance the ball is made of, moving from the source to the receiver (we are therefore in case B above).

(2) **Water-pump-motor.** *S* is a hydraulic gear pump, *R* is a hydraulic motor, *M* is the water that flows in a closed circuit from the pump to the motor and *I* is momentum. More precisely, through

the rotation of its gears, the hydraulic pump communicates momentum to the water, putting it into circulation in the pipes. The (high pressure) water flowing in the pipes confers part of its momentum, and therefore of its energy, to the motor gears, which are thus set in motion.

The water is therefore the material carrier of momentum, and momentum is the immaterial carrier of energy.

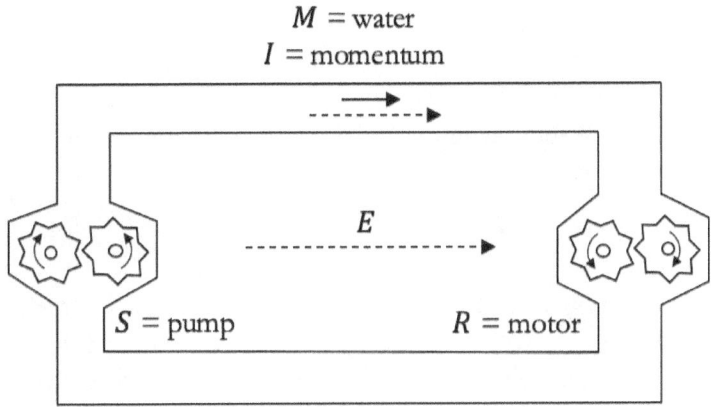

Figure 3 The *energy flow diagram* schematically describing a process where a pump transfers energy to a hydraulic motor, using for this the material carrier "water" and the immaterial carrier "momentum."

Also in this example, as in the previous one, besides the presence of a current of energy and momentum (two immaterial substances), we have the presence of a current of matter (water under pressure) moving in a closed circuit, from the source to the receiver, and back (we are therefore again in case B).

(3) ***Boiler-water-heater.*** S is a boiler, R is a heater, M is the hot water flowing from the boiler to the heater, and back abd I is entropy. More precisely, S communicates *entropy* to the water, heating it up, by putting it in contact with a vessel at high temperature (entropy passes spontaneously from regions of higher temperature to regions of lower temperature). Via a pump, the hot water circulates in the pipes and reaches the radiator, to which it transfers part of its entropy, by contact (cooling down).

The water is therefore the material carrier of the entropy, and entropy is the immaterial carrier of the energy.

Also in this example, as in the previous two, besides the presence of a current of energy and entropy (two immaterial substances), there is a current of matter (hot water) moving in a closed circuit, from the source to the receiver, and back (we are therefore in case B).

Note: It is of course also possible to transfer energy from the boiler to the radiator without letting the water circulate. In this case, however, the intensity of the entropy current from the boiler to the heater is going to be much lower and consequently the efficiency of energy transfer will also be reduced. Such a situation corresponds to case C, since there wouldn't be then a significant transfer of matter (the water doesn't flow).

(4) ***Battery-electricity-bulb.*** S is an electrical battery, R is a light bulb, M is the electricity (i.e., the current of electrons moving along the wires) and I is momentum and the electric charge. More precisely, through the electromotive force, S communicates momentum to the negatively charged electrons, which are then "pushed" from the negative to the positive pole of the battery.[3] Arriving in the bulb, where there is a strong resistance, due to the friction entropy is created (from scratch). This means that the bulb works as a *transceiver:* energy goes into the bulb carried by electricity, and part of this energy is transferred to the bulb by means of the entropy produced (energy which in turn the bulb will transfer to the surrounding environment by means of an electromagnetic perturbation, called light).

The electrons are therefore the material carriers of momentum and electric charge, whereas momentum and electric charge are the immaterial carriers of energy.

Also in this example, as in the previous ones, in addition to the current of energy, momentum and electric charge (three immaterial substances), we also have the presence of a current of matter (the electrons) which flows from the region of higher to the region of lower electric potential (we are therefore in case B).

Note: the light bulb, as we have seen, is an *energy transceiver.* An

---

[3] The direction conventionally indicated for the electric current is that of positive charges, so opposite to the actual direction of motion of the electrons in a conductor.

energy transceiver, in general, is an entity that receives energy through a specific carrier and transfers it through a different one. The bulb, as an electrical resistance, transfers energy from the carrier "momentum" to the carrier "entropy" (internal combustion engines do exactly the opposite).

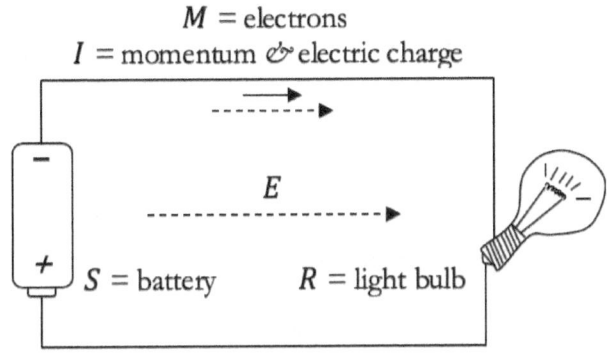

Figure 4 The *energy flow diagram* schematically describing a process where an electrical battery transfers energy to a light bulb, using for this the material carrier "electricity" and the two immaterial carriers "momentum" and "electric charge."

(5) ***Tank-gasoline-motor.*** S is the tank of a car, R is the motor of the car, M is gasoline. More precisely, thanks to a pump, the material substance "gasoline" is conveyed to the motor. Inside the motor, a chemical reaction takes place (combustion): the gasoline is combined with oxygen and produces a large amount of entropy, thus transferring energy to the pistons, which receive a boost (i.e., momentum).

Gasoline is therefore the material carrier of energy and there is no immaterial carrier in this case.

It is indeed gasoline itself that, by combining with the oxygen molecules in the piston, is destroyed, and in the reaction conveys energy to the substances produced (carbon dioxide and water), which thus acquire a large momentum, part of which is transferred to the piston.[4] In other words, in this case the only energy carrier is the

---

[4] Momentum is not for this created from nothing. The total momentum, given by the (vector) sum of momentum of the different products of the combustion

quantity of substance "gasoline" (we are therefore in case A).

In the five examples described above, in combination with the immaterial current of energy a current of material substance is always present, of non-zero intensity: bowling ball, water, electrons, gasoline (cases A and B). In some of these examples, but not in others, the material carrier of energy flows in a closed loop.

The distinction between "closed circuit" or "open circuit" systems, however, does not have any deep physical meaning. It is simply about observing that different configurations are possible. On the other hand, this distinction highlights an important point:

Energy doesn't necessarily flow along with the material carrier.

This was already quite clear in Examples 2, 3 and 4, considering that the material carrier, in contrast to energy, moves along a closed circuit. In Example 2, the material carrier departs from the source as "high pressure" water, and once having transferred energy to the receiver, makes its way back to the source as "low pressure" water. In Example 3, the material carrier leaves the source as "high temperature" water, and once having transferred energy to the receiver, makes its way back to the source as "low temperature" water. In Example 4, the material carrier quits the source as "high potential" electric current, and once having transferred energy to the receiver, makes its way back to the source as "low potential" electric current. In other words, energy can flow regardless of the flowing of its material carriers.

I have already emphasized this fact in Example 3, observing that the immaterial substances "entropy" and "energy" can flow from a boiler to a radiator even when the water is not circulating in the pipes.[5] Let me now analyze this possibility in more detail, in more specific examples.

(6) **Newton's pendulum.** *S* is the metal ball at the extreme left of

---

reaction, is obviously equal to the total momentum possessed by gasoline and oxygen prior to the reaction.

[5] More simply, one can bring a small metal object to the flame of a candle: very soon a flow of entropy and energy will pass through the object, reaching the fingertips and activating one's nociceptors.

the pendulum, R is the metal ball at the extreme right of it, M consists of the metal balls in-between them and I is momentum. More precisely, S gives M its energy, by transferring all its momentum to the second ball, which then transfers it to the third, then the third to the fourth, and so on, until the last ball, which is the receiver, is placed in motion, and therefore receives energy. As is known, the process takes place without any of the intermediate balls (forming the substance of the material carrier) moving.

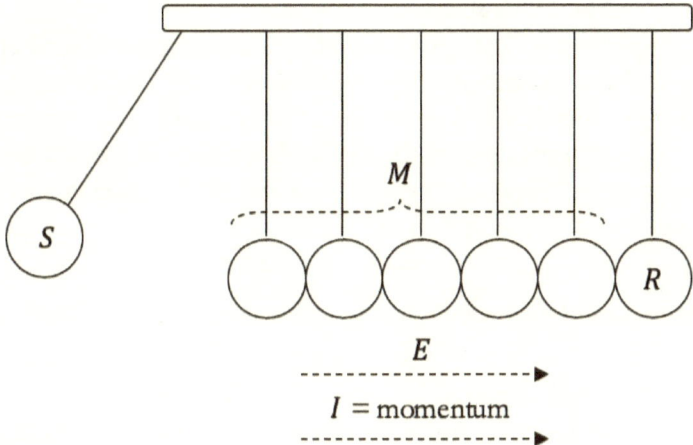

Figure 5 The *energy flow diagram* schematically describing a process where a metal ball transfers energy to another metal ball, using a number of intermediary metal balls as the material carrier, and momentum as the immaterial carrier.

In other words, M doesn't flow: the metal balls remain steady (the velocity of their center of mass is zero). However, what instead does propagate is a (shock) *wave of longitudinal compression and decompression* between the metal balls (which are elastic), i.e., a deformation (the effect is similar to the well-known "domino effect").

The intermediate balls are the material carrier of momentum, and momentum is the immaterial carrier of energy.

However, the material carrier doesn't flow together with the immaterial carrier (we are therefore in case C).

(7) **Sound waves.** The situation with Newton's pendulum is very similar to what happens when a sound wave propagates in the air

(M), for example between the speaker of a radio (S) that generates the wave and the eardrum of an ear (R) that receives it.

As for Newton's pendulum, a sound wave is also a longitudinal perturbation, that is, a wave of compression and decompression of the air molecules, which propagates without the need of any material current: the air molecules are set locally in motion by the oscillation of the speaker and communicate, always locally, through collisions, their movement to the closest molecules, and so on, until the oscillation reaches the tympanic membrane of the ear, which is also put in oscillatory motion, thus receiving energy.

It is important to distinguish the situation of the transfer of energy through the propagation of a sound wave with that of an energy transfer produced, for example, by a hot air heater. The carrier is always air, but in the case of the hot air heater it is driven by the fan. In this way, a "wind of matter" is created, that is, a current of the material carrier (to which is associated an immaterial entropy current), which is not the case for the sound wave.

(8) **Hand-trolley-rope.** S is a hand, R is a trolley (that you want to pull), M is a rope that on one side is held by your hand and on the other side is tied to the trolley and I is momentum.

This example is perhaps even more significant in illustrating the fact that the material carrier doesn't have to flow from the source to the receiver, for energy to be transported. In fact, the person transfers energy to the trolley by *pulling the rope*. Obviously, the stuff the rope is made of doesn't flow from the hands of the person to the trolley. The rope is simply placed under tension.

One usually says in this case that a *force* is applied, but force, as evidenced by Newton's second law, is nothing but an expression of a *current of momentum.*[6]

The rope is the material carrier of momentum, and momentum is the immaterial carrier of energy.

However, the material carrier doesn't flow together with the immaterial carrier (we are therefore in case C).

---

[6] According to Newton's second law: $f = dp/dt$.

5 Energy transfer between two intraphysical consciousnesses

I would like now to describe the process of energy transfer between an intraphysical human operator (S), who exteriorizes energy, for instance through his/her *palmochakra*, and another intraphysical consciousness (R), able to receive it. Here of course I'm assuming that the process is completely objective, i.e., that S and R are not simply imagining exteriorizing and receiving energy. In other words, it is not in question here the fact that there would be an objective energy current between S and R.

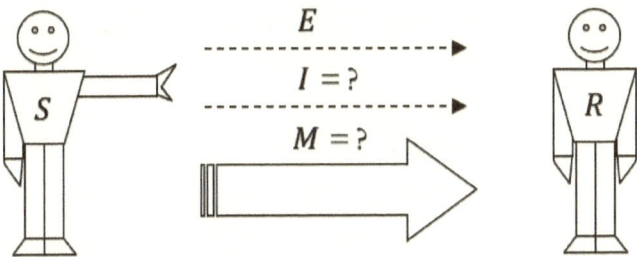

Figure 6 The *Energy flow diagram* schematically describing a process where an intraphysical consciousness transfers energy to another intraphysical consciousness, through some unknown carriers.

Based on what has been described in the previous sections, we are now able to formulate some conceptually well-posed questions, about the nature of this process of emission and absorption of energy between two intraphysical consciousnesses. Since energy, to be transported, requires the presence of a material carrier, the first question one must ask is the following:

What is the nature of the material carrier M that connects S and R, allowing for the transfer of energy?

Also, it is necessary to ask if the material substance is present in the environment between S and R, or if it is exteriorized by the intraphysical consciousness S, or perhaps by both consciousnesses.

One possible hypothesis is that around S and R, a *field of "subtle" matter* is always present, with a given spatial range, and that when S and R are sufficiently close in spatial terms, their "subtle" matter fields are able to co-penetrate, thus forming the material carrier M that will then allow for the passage of energy. But whether M is exteriorized at the moment by S and/or R, whether it is already present in the environment, or accompanies S and R as a kind of "personal atmosphere of subtle matter," it is also important to clarify if in the energy transfer process between S and R the material carrier M does flow or not. In other words, the second question one must ask is:

Does the material carrier M flow from S to R, or is its current zero?

Depending on the response, there can be the following two additional questions:

If the material current is zero, what is the nature of the immaterial substances carrying energy?

If the material current is non-zero, is energy only carried by the material substance, or also by the immaterial ones? And what is their nature?

It is of course not easy to provide answers to the above questions. Nevertheless, if the energy transfer between the intraphysical consciousnesses S and R is objective, then we know that, necessarily, a material substance M must exist (however "subtle" it is), which is the material carrier of the immaterial substance 'energy'.

The central question is therefore about clarifying whether the exteriorization of energy on the part of S is mainly due to an exteriorization of a material substance (as when, with our lungs, air is blown outwards from the mouth), or if it is more like a local perturbation propagating in space, carried by an immaterial substance (such as momentum), without there being any transport of matter (such as when we emit a sound wave with our vocal chords); or if both mechanisms take place simultaneously.

Clearly, these same questions can be addressed (and may receive very different responses) when the psycho-energetic process possibly involve even more "subtle" substances, such as in the case of empathic and telepathic communications.

AutoRicerca, Issue 20, Year 2020, Pages 233-265

6 Energy transfer within an intraphysical consciousness (the VELO technique)

Of course, we can ask questions similar to the previous ones also with regard to the internal movements of energy, as in the methodology called VELO (Alegretti, 2008; Trivellato, 2008). In this case, the material carrier would be the very stuff that forms our extraphysical body, usually called energosoma (or energetic body, holochakra, etheric body, pranamaya kosha, etc.). However, the energosoma is not necessarily a homogeneous structure and when during the execution of the VELO technique one produces (according to a specific protocol) an alternating energy current along its structure, it is necessary to ask whether the phenomenon in question can be understood as the movement of a material fluid substance (belonging or not to the energosoma), which would pass through a more rigid structure (as for example in the case of the fluid 'air', in the physiological breathing, or the fluid 'blood', in the internal blood circulation), or whether it is rather a propagation of energy only carried out by immaterial carriers. In the latter case, the energosoma would merely constitute the material stuff supporting the conduction of the immaterial carriers, which would longitudinally transport energy along the body, but with no transport of matter.

Of course, both these possibilities could be implemented simultaneously, during the execution of the technique. That is, a material fluid could pass through the structure of the energosoma and, at the same time (or as a result of this material movement), a current of immaterial carriers (e.g., momentum) would also flow through it, in a process that could be similar to that of the propagation of a sound or electromagnetic wave.

I will not advance myself further into these considerations, since the purpose of this paper is not to clarify the specific nature of these phenomena, but rather to suggest a clear conceptual language for addressing them, so as to formulate well posed questions.

7 Energy or forms of energy?

It is important at this point to clarify an important misconception. It is customary, even by physicists, to distinguish between different "forms of energy." This subdivision of the immaterial substance "energy" in different forms, however, is quite misleading and should be avoided as much as possible. Let me explain.

Historically, the distinction between different forms of energy has followed two main criteria: (1) how the energy can be stored, that is, contained in a physical system; (2) how the energy can be exchanged between different physical systems. The first criterion has led to the distinction between forms of energy such as kinetic energy, potential energy, elastic energy, internal energy, etc. The second criterion has instead given rise to the distinction between forms of energy such as heat, work, electricity, chemical energy, etc.

The first criterion is usually applied when the system under consideration can be divided into subsystems.[7] Let me consider a simple example: a small body of mass m (a classical particle) free falling in Earth's gravitational field, near the surface. In this case, it is customary to say that the body has a certain amount of *kinetic energy* K (carried by its momentum p, according to the well-known formula: $K = p^2/2m$) and a certain amount of *gravitational potential energy* V (given by the formula $V = m \cdot g \cdot z$, where g is the acceleration of Earth's gravitational field and z the height of the body).

This description is however conceptually incorrect. In fact, the gravitational potential energy V is not owned by the body, but by the physical entity named "gravitational field" in which the body is immersed. What happens when the body falls, is that the gravitational field transfers some of its energy to the body, by exchanging momentum (conversely, the gravitational field, as a container, receives back energy from a body, when it is lifted up).

---

[7] Mathematically speaking, this statement means that the Hamiltonian function describing energy can be written as a sum of independent terms, in the sense that each term depends on variables that do not appear in the other terms.

In other words, there is not a "kinetic energy" and a "potential energy"[8] that is, two different "forms of energy," jointly possessed by a body of mass m. On the contrary, there is a single immaterial substance, simply called "energy," owned by two different physical systems: the material body and the gravitational field.

The other criterion usually applied to distinguish between different forms of energy is in relation to the modalities of its transfer. Usually, we say that energy is transferred from one system to another in the form of heat, work, chemical energy, electrical energy, etc. These forms, however, as we have seen, have nothing to do with energy itself, but rather with its carriers.

Once the distinction between the concept of "energy" and the concept of "energy carrier" is clear, it also becomes clear that only a single form of energy exists, although it may be carried from one system to another with entirely different modalities. What one can observe, as shown in different examples above, is that *energy always flows with at least another substance* (understood here as an extensive quantity), which can be either material or immaterial. These substances that accompany the flow of energy are its carriers. Carriers may change, of course, but this doesn't mean that the form of energy changes. Indeed, only its mode of transportation changes. To use a metaphor, surely it makes no sense to speak of different "forms of eggs," distinguishing "car-eggs" from "bicycle-eggs," depending on the vehicle with which they are transported.

This means that so-called *non-homogeneous* (or hybrid) *transducers*, such as electromechanical, electro-optical, magnetoelectric, piezoelectric transducers, etc., should not be considered as devices where the energy input would be of a "different form" than the energy output. Energy is one and only one! However, what the transducer does, is to change the type of carrier that transports energy.

If we follow this logic, we can see that speaking of different "forms of energy" is as inappropriate as speaking of different "forms of electric charge," depending on whether it is carried by electrons, protons, muons, etc., or different "forms of

---

[8] The term "potential energy," on the other hand, can be considered correct if it is understood not as a form of energy possessed by the body of mass m, but as the energy it could possibly (potentially) receive from the gravitational field in which is immersed.

momentum." Just as it makes no sense to speak of a "bowling ball form of momentum," but it makes sense to distinguish the immaterial substance "momentum" from its material carrier "bowling ball," in the same way it makes no sense to confuse the immaterial substance "energy" and its possible carriers, whether material or immaterial.

8 Mass, energy and matter

I have already mentioned the importance of distinguishing the concept of *matter* from that of *energy*, since the two concepts are based on entirely different ontological categories, energy being a *property of matter*, whose characteristic is to behave like a substance (*substance-like* property). Similarly, I mentioned the importance of not confusing the concept of *mass* with that of *matter*. According to the theory of relativity (special and general), we know that (until proven to the contrary) *mass and energy are two entirely equivalent ways of talking about a same reality*: it is exactly the same concept, only described with different units of measurement.

Energy, like mass, determines the intensity with which a physical entity (possessing such energy) is able to receive additional energy from a gravitational field, by means of the weight-force (which, like all forces, describes the intensity and direction of a momentum current). And energy, like mass, also determines the resistance exerted by a physical entity in altering its state of motion (inertia). In other words, energy and mass have same characteristics and, consequently, describe the same physical property. We can therefore speak indifferently of mass, energy, or mass-energy (a redundant term). Thus, as it is necessary to distinguish between matter and energy, it is also necessary to distinguish between matter and mass, the latter being a substance-like property of matter, equivalent to energy.

The main difficulty in distinguishing matter and mass (and therefore matter and energy) lies in the fact that in physics mass was initially understood as "quantity of matter." But of course, we must not confuse the amount of a given material substance, for instance expressed by the number of elementary entities of a given

AutoRicerca, Issue 20, Year 2020, Pages 233-265

kind present in a system, with the mass (or energy) carried by those entities.

9 Terminological problems

Considering the conceptual clarification offered in the previous sections, and especially the fundamental distinction between the concept of "energy," which cannot be decomposed into distinct forms, and the concept of "energy carriers," which on the contrary are numerous and are certainly to be distinguished, we can question the relevance of terms such as: energies (in the plural), "subtle" energies, extraphysical energies, immanent energies, consciential energies, energetic dimension, energosoma, energetic body, etc.

First, let me observe that using the term "energy" in the plural, that is, to speak of "energies," is obviously misleading, as this would suggest the existence of more than one immaterial substance associated with the concept of energy. Instead, as I have emphasized, there is a single immaterial substance called "energy," which has the remarkable property of being conserved (unless evidence to the contrary) in all processes of interaction between physical entities (in the sense that it can be neither created nor destroyed). Therefore, it is desirable to avoid stating the term "energy" in the plural.

Another inaccuracy is to qualify the term energy, for example when one says "subtle energy," or even worse "subtle energies." If the immaterial substance "energy" is unique, it is obviously incorrect, as we have explained above, to distinguish between different forms of it. It is incorrect to do so when dealing with ordinary physical systems (although it is common practice among physicists, including the author) and a fortiori it is also incorrect to do so when one describes non-ordinary physical systems. Of course, it is not here in question the utility of using terms such as "subtle" to identify the non-ordinary nature of the phenomenon under consideration. The point is that this adjective doesn't refer to energy, but to the material carriers of energy.

In other words, if the term "subtle energies" is understood as an abbreviation which stands for "energy conveyed by subtle material substances," its use is certainly acceptable. However, in my experience, when one uses this expression, this is not the way it is usually understood. Therefore, my advice is to use as much as possible the more appropriate terms of "subtle matters," "subtle substances," or "subtle fluids," instead of "subtle energies." Obviously, the same observation applies to the other terms mentioned above. Instead of speaking of "immanent energies," it would be preferable to say "immanent matters," or "immanent substances." Same thing for the term "consciential energies," which should preferably be replaced by "consciential matters," or "consciential substances."

The term "energosoma" (or "energetic body") also lends itself to possible misunderstanding, since each vehicle of manifestation possesses energy, and is therefore an energosoma! Every physical entity has, until proven to the contrary, energy, and to emphasize that a specific entity is of an energetic nature is a kind of pleonasm, which is likely more to confuse than to clarify.

Let me observe that the term "energy," associated with "soma," is usually used to indicate the more *fluid* and *translucent* nature of this interface, when compared to the more solid in appearance and opaque somatic vehicle, also because in one's imagination it is customary to associate the concept of energy with something fluid, vibrant, luminous, electric, etc. In that sense, it would be preferable to use terms such as "fluidosoma," or "vibrational body."

The concept of *thosene*, as defined in *conscientiology* (Vieira, 2002), could also be revisited, replacing the "e" (or "ene") of "energy," with a "ma," referring to the "matter" aspect, constitutive of physical entities, be them ordinary or non-ordinary. That is, "thosema" instead of "thosene." Indeed, matter, or rather matters, are the founding elements giving rise to the structures supporting our cognitive processes, such as emotions and thoughts. And as there are matters of different nature, more or less "subtle," so there are emotional and mental processes of different nature, depending on the substances (and corresponding structures) which are carrying them. For example, we can feel emotions and think using primarily the matters of our soma, or do the same using the paramatters of our psychosoma, when in extracorporeal states, or the metamatters of our mentalsoma, for example in a mentalsomatic projection.

Let me consider now the term "extraphysical." Here, depending on the meaning one attaches to the prefix "extra," the understanding of the term may vary. First, it is good to understand the etymology of the word "physical." It can be related to the Greek word "phusis," which means "that which is put into existence," which in turn derives from the Greek verb "phuoo," which means "to create, springing up." More customarily, the word is associated to the (always Greek) term "physis," which means "nature," to be understood as "the world," that is, "that which exists in a substantial sense." In short, whatever the way one wants to understand the etymology of the word, this certainly doesn't create a separation between "coarse-grained" and "fine-grained" realities, but potentially encompasses all of reality.

The term "extraphysical" is therefore not to be understood as that which goes beyond the physical, as this would make no sense. The prefix "extra" is rather to be understood in the sense of "extra-ordinary," that is, "non-ordinary." This means that the vehicle connecting the soma and the psychosoma – the *fluidosoma* – is to be considered as an "extraphysical vehicle," in the sense of being a "non-ordinary physical vehicle." Therefore, when we talk about "extraphysical dimensions," what we have to understand by this term is "non-ordinary physical dimensions," i.e., "physical dimensions formed by non-ordinary material substances."[9] On the other hand, terms like "extraphysical energies" should be avoided altogether, and be replaced by "extraphysical matters," to be understood as "non-ordinary physical matters."

10 Energy and data

The concept of energy is obviously not the only relevant concept when studying the physical properties of substances, be them ordinary or non-ordinary, living or nonliving. Another concept of undoubted importance is that related to the *information* that is

---

[9] Similarly, our "intraphysical" condition denotes not so much a condition of physicality, but rather a condition of ordinary physicality, associated to our experience of an ordinary, three-dimensional physical space, inhabited by classical substances of a specific kind.

constantly being exchanged between different physical systems. Many self-researchers affirm, and rightly so, that what most characterizes psycho-energetic phenomena is not so much the amount of energy which is being exchanged, as the (meaningful) information that is conveyed in this way. Undoubtedly, the psycho-energetic phenomena human beings are able to manifest, both in their intraphysical and extraphysical conditions, do sponsor dynamics that in addition of exchanging energy, also exchange information. In other words, the aspect of "communication of a meaning" in these energy exchanges is probably of primary importance, if one wants to understand the true nature of these phenomena.

A good analogy is that of the human oral language. Being interested in the energetic aspect of human oral communication is certainly important and requires a thorough knowledge of the characteristics of vocal cords, tympanic membranes, and waves propagating in the air. But to think to understand what really happens when two humans verbally communicate, taking into account only the energetic aspect of the communication, it would obviously be totally inadequate. In other words, to fully understand the interactions happening in the ambit of a human conversations, one must also, and above all, be interested in syntax and semantics, i.e., in the structure of the language and the meaning conveyed by such a structure, as well as, of course, in the way this meaning is modulated according to the different contexts and minds participating in the interaction.

On the other hand, it is also true that to speak one needs to have well-functioning vocal cords, and to listen one needs well-functioning ears. Moreover, under water, it is certainly not very practical to have a clear conversation with one's interlocutor (regardless of respiratory problems). I say this just to draw attention to the fact that one thing is a flow of data, and the information it potentially carries, and another thing is its transportation.

When I was writing this article, because of a small earthquake, an unexpected power blackout occurred, which suddenly wiped out the flow of energy entering my computer. The consequence of this small energetic incident is that the whole document I was working on was destroyed, together with the data it contained. With this anecdote I just want to draw attention to a simple fact: the transport of data, and of the information associated with those data, takes energy, and transport of energy, as we have seen, requires at least the presence of one material carrier.

So, even though I do certainly agree that the exchanges of energy associated to mental communications need to be understood not only in terms of quantity of energy, currents of energy and currents of energy's carriers, but also in terms of content, relation, meaning, coherence, structure, etc., it is important to always remember that every communication needs, to be implemented, the presence of material and immaterial substances, able to sustain it. Therefore, the understanding of such exchanges cannot totally disregard the understanding of the nature of the substances that carry the data, which are the same as those carrying the energy. In this regard, I observe – and on this I conclude my brief parenthesis on "information" – that also the physical quantity "amount of data" behaves like an immaterial substance, to which one can associate a specific current, the intensity of which is usually calculated in bits per second.

11 Classical and non-classical substances

Before concluding this article, some words should be said about the subject of *non-classical material substances*. I have assumed in this paper, not to complicate the discussion, that the matters at stake, both in ordinary and non-ordinary physical systems, were classical, in the sense of being present in each moment in our *three-dimensional* ordinary physical space (OPS). But this is certainly not the rule. A typical example of a non-classical material substance is the carrier of electromagnetic waves. In the past the carrier was named by physicists the *ether*, but with the advent of Einstein's relativity the term almost totally disappeared from their vocabulary.

This is because, as a consequence of the theory of relativity, it seemed totally impossible to attribute to the ether a specific state of motion in space. And if the ether did not have a proper motion, the logical consequence for many physicists was to simply decree its inexistence, based on the prejudice that our three-dimensional space would be the theater containing the totality of all that exists, and that all spatial entities were compelled to possess well-defined states of motion.

But once the ether has been eliminated, the electromagnetic waves became, all of a sudden, very paradoxical perturbations, able to propagate into the *nothingness*, i.e., without the presence of a material carrier capable of supporting their propagation. Actually, if the concept of the ether went out the front door, it re-entered from the back-door, in a different guise. Physicists today no longer speak of the ether, this is true, but they speak of the *vacuum* and its properties, distinguishing this concept from the one of *nothingness*; or they speak of *fields*, meaning with this the set of properties characterizing specific regions of the three-dimensional space.

But the trick to delete the word "ether" does not solve the problem of determining what the physical vacuum or the physical fields are. There is no doubt that since they possess physical properties, they must be physical entities, made of some material substances. But these substances, although material, are certainly not of the ordinary kind. Indeed, the impossibility to describe them in terms of specific states of motion suggests that they are matters that do not belong to our three-dimensional OPS. And if this is what is indicated by relativity theory, the situation becomes even more serious when quantum theory comes in.

Indeed, it is known that quantum entities, while being certainly physical, cannot in any way be described as substances that would sojourn permanently in our OPS, their spatiality being very different from that of the objects of our everyday experience. Obviously, I cannot go here into the details about these issues, which are conceptually quite subtle. To deepen their understanding, I recommend the reading of the works of the Belgian physicist *Diederik Aerts*, especially (Aerts, 1990, 1999). Some references to the work of Aerts can also be found in some of my writings published in this journal (Sassoli de Bianchi, 2006a,b, 2009b), or in some of my most recent publications (Sassoli de Bianchi, 2011a, 2012, 2013a,b).[10]

In particular, in (Sassoli de Bianchi, 2013a) I suggest looking at our *physical space* as an entity that extends beyond the simple three-

---

[10] For even more recent publications by the author, the reader can consult his homepage at: *www.massimilianosassolidebianchi.ch*.

AutoRicerca, Issue 20, Year 2020, Pages 233-265

dimensional theater of our ordinary experience. The classical substances, that in every moment possess a well-defined position and momentum, are those that by definition stably reside in the *three-dimensional* OPS, that we all know, but this space is in turn contained in larger theatres, of an *extra-ordinary* nature, and it is in these non-ordinary ambits that usually the quantum entities stay.

These larger spaces, even if extra-ordinary from the viewpoint of our ordinary perception, are still always part of the physical space, since, as I have many times emphasized, everything that exists has by definition some kind of physicality, that is, of materiality.

Therefore, the conceptual framework presented in this paper remains in principle also valid for quantum material substances, although their way of behaving and manifesting their presence differ from that of classical macroscopic bodies. Their presence in the three-dimensional space is in fact only potential: they are available in being "sucked up" into it, in certain circumstances, and their availability can be quantified by means of probabilities, but their primary place of residence is not the OPS of our three-dimensional intraphysical experience (a fact usually described in the scientific literature by means of the concept of *non-locality*).

The same is undoubtedly true also for the subtler paramatters, although their characteristics are probably very different in comparison to the quantum entities today studied by physicists, as are probably also very different the extra-ordinary spaces in which these paramatters usually reside.[11] But regardless of the nature of the various material substances and the spaces (ordinary or non-ordinary) in which they usually reside, I think there are no reasons to deny one of the basic principles outlined in this paper, namely that to transport energy between two entities, regardless of their nature and spatiality, the presence of at least one material carrier (ordinary or non-ordinary) and of possible additional immaterial ones, is always required.

---

[11] Most likely, the distinction between classical and quantum behaviors does not apply only to the matters today studied by conventional physics, but also to the paramatters that form the higher vehicles of manifestation and related existential dimensions of the consciousness.

12 Conclusion

I conclude this article with a few remarks. Regarding the issue of the inadequacy of the concept of "energy form," one could argue that also in conventional science the confusion between energy and forms of energy is continually promoted. This is certainly true, but it is not because a misconception is promoted by a majority that this justifies its perpetuation. Furthermore, I believe that in the study of paramatters it is particularly important to emphasize the non-subjectivity of such realities – as for example the different vehicles and corresponding interfaces that form the holosoma of the being-consciousness – thus not referring to them as "energy structures" but rather as "material structures."

Much of the ideas expressed in this article are inspired by the work of *Karlsruhe's German school of physics* (Falk et al, 1983; Schmid, 1984; Herrmann, 2000). In this school, however, the concept of "immaterial substance" is not considered, in the sense that a distinction between "material carriers" and "immaterial carriers" is not made (one speaks of energy carriers in general terms, whatever their nature).

It is important to observe that the immaterial character of a substance is such because it is a *substance-like property* that cannot exist without the support of a material substance. That this supporting substance is subtle or not, that is not the point. It is therefore good practice not to confuse the "subtle" paramaterial substances with the immaterial substances associated with them, such as for example energy. Energy is, until proven to the contrary, a purely immaterial quantity, regardless of the spatial, dimensional and existential context in which it is considered.

Appendix

In this appendix I will only provide a fundamental relation between *intensive* and *extensive* quantities, which determines the intensity of a current of energy $I_E = P$ (power). This relation highlights the fact that to every carrier of energy (characterized by an extensive quantity, like the amount of matter, the electric charge, momentum, entropy, etc.) is associated a specific intensive quantity (chemical potential, electric potential, velocity, temperature, etc.) that quantify how much the carrier is charged with energy or, better, the "thrust" the carrier receives, which determines the intensity of the energy flow (Falk et al, 1983; Schmid, 1984). More precisely, we have the following relation:

$$I_E = \mu \cdot I_M + \phi \cdot I_Q + v \cdot I_p + T \cdot I_S + \cdots$$

where I_M is the current intensity of the material carrier (measured in number of moles per second) and μ is the chemical potential, I_Q is the current intensity of the immaterial carrier "electric charge" (measured in amperes, that is, in coulombs per second) and ϕ is the electrical potential, I_p is the current intensity of the immaterial carrier "momentum" (measured in newtons, i.e., in huygens per second, usually associated with the concept of force) and v is the speed, and I_S is the current intensity of the immaterial carrier "entropy" (measured in carnot per second) and T is the absolute temperature.

Bibliography

Abs de Lima, André (2005). An Analysis of Bionergy as studied by Projectiology and other Conventional Sciences, Journal of Consciousness, Volume 7, No. 27, pp. 255-268.

Aerts, Diederik (1990). An attempt to imagine parts of the reality of the microworld, in: *Problems in Quantum Physics II; Gdansk '89*, eds. Mizerski, J., et al., World Scientific Publishing Company, Singapore, pp. 3-25. (Italian translation: AutoRicerca 2, 2011).

Aerts, Diederik (1999). The Stuff the World is Made of: Physics and Reality, in: *The White Book of "Einstein Meets Magritte"*, eds. Diederik Aerts, Jan Broekaert and Ernest Mathijs, Kluwer Academic Publishers, Dordrecht, pp.129-183.

Alegretti, Wagner (2008). An Approach to the Research of the Vibrational State through the Study of Brain Activity, Journal of Consciousness, Vol. 11, No. 42, p. 217. (Also published in this volume).

Bruce, Robert (2007). *Energy Work*, Hampton Roads Publishing Company.

Falk, G., Herrmann, F. & Schmid, G.B. (1983). Energy forms or energy carriers? Am. J. Phys. 51, pp. 1074-1077.

Herrmann, F. & Job, G. (1996). The historical burden on scientific knowledge, Eur. J. Phys. 17, pp. 159-163.

Herrmann, F. (2000). The Karlsruhe Physics Course, Eur. J. Phys. 21, pp. 49-58.

Sassoli de Bianchi, Massimiliano (2006a). A Dialogue About Science, Reality and the Consciousness – Part I, Journal of Consciousness; Volume 9, No. 33, pp. 365-418. Also published (both in English and in Italian) in: *Science, Reality & Consciousness*, AutoRicerca 7 (2014).

Sassoli de Bianchi, Massimiliano (2006b). A Dialogue About Science, Reality and the Consciousness – Part II, Journal of Consciousness, Volume 9, No. 34, pp. 3-56. Also published (both in English and in Italian) in: *Science, Reality & Consciousness*, AutoRicerca 7 (2014).

Sassoli de Bianchi, Massimiliano (2009a). Interdimensional energy transfer: a simple mass model, Journal of Consciousness, Volume 11, No. 43, pp. 297-315. (Also published in this volume).

Sassoli de Bianchi, Massimiliano (2009b). Reply to David Lindsay's Letter [Journal of Consciousness, Vol. 12, No 45, July 2009], Journal of Consciousness, Volume 12, No. 45, pp. 65-72.

Sassoli de Bianchi, Massimiliano (2011a). Ephemeral Properties and the Illusion of Microscopic Particles, Found. Science, Volume 16, Issue 4, pp. 393-409.

Sassoli de Bianchi, Massimiliano (2012). From permanence to total availability: a quantum conceptual upgrade, Found. Science, Volume 17, Issue 3, pp. 223-244.

Sassoli de Bianchi, Massimiliano (2013a). The δ-quantum machine, the k-model, and the non-ordinary spatiality of quantum entities, Found. Science, Volume 18, Issue 1, pp 11–41.

Sassoli de Bianchi, Massimiliano (2013b). The observer effect, Found. Science, Volume 18, Issue 2, pp 213–243.

Schmid, G.B. (1984). An up-to-date approach to physics, Am. J. Phys. 52, pp. 794-799.

Tiller, William A. (1993). What Are Subtle Energies? Journal of Scientific Exploration, Vol. 7, No. 3, pp. 293-304.

Trivellato, Nanci (2008). Measurable Attributes of the Vibrational State Technique, Journal of Consciousness, Vol. 11, No. 42, p. 165. (Also published in this volume).

Vernon Vugman, Ney (1999). Conscientiology and Physics: A Desirable Couple? Journal of Consciousness; Volume 1, No. 4, p. 289.

Vieira, Waldo (2002). *Projectiology, A Panorama of Experiences of the Consciousness outside the Human Body*, Rio de Janeiro, RJ – Brazil, International Institute of Projectiology and Conscientiology.

Zamperini, Roberto (1998). *Energie sottili*, Macro Edizioni.

AutoRicerca, Issue 20, Year 2020, Pages 233-265

Note: This article was previously published in: Journal of Consciousness 16, 2013, pp. 9-40, with the title: "Subtle energies or subtle matters? A conceptual clarification." An Italian translation is also available in: *Energia*, AutoRicerca, Numero 6, Anno 2013, pp. 51-87.

AutoRicerca

A heuristic density model to explain the gap between physical and extraphysical dimensions

Massimiliano Sassoli de Bianchi

Issue 20
Year 2020
Pages 267-287

Abstract

A simple density model is presented to explain the energetic separation between the physical and extraphysical dimensions and the possible role of the energosoma as a mediator structure of variable density, able to increase the efficiency of an interdimensional energy transfer. The model, which is only heuristic, is discussed both from the viewpoint of classical and quantum physics.

AutoRicerca, Issue 20, Year 2020, Pages 267-287

1 Introduction

Our multidimensional reality is made of different energetic substances. Physical substances, called *matter*, form our physical (material) dimension, which has been thoroughly investigated by *physicists*, particularly in the last two centuries. Extraphysical substances, that we may also call *paramatter*, are instead assumed to make up our much vaster extraphysical dimensions and are the domain of investigation of the *paraphysicist*, an emerging scientific figure that we can expect to gain more recognition in the future. However, contrary to physics, paraphysics has not yet reached on this planet the level of development of a quantitative, fully mathematized hard science, and intraphysical (incarnated) paraphysicists are in the same situation today as were Greek philosophers like Democritus (about 460-370 B.C.) when speculating about the atomic or non-atomic nature of physical matter.[1]

A strategy paraphysicists can certainly adopt for the time being, to study the properties of paramatter and its interactions with ordinary matter, is to exploit all possible analogies with what is already established regarding our physical dimension. Indeed, it is to be expected that some of the general principles and models that have so far been developed in physics will also prove their usefulness, *mutatis mutandis*, in the understanding of our extraphysical reality, at least in the initial stages of development of paraphysics. Of course, this exercise has to be carried out *cum grano salis*, otherwise, as emphasized for instance by Vernon Vugman (1999), migration of concepts from physics to paraphysics may result in some form of reductionism and possibly delay the development of the latter.

The purpose of the present article is to discuss some simple heuristic models in order to gain a better understanding of the possible interaction mechanisms between the substances forming different existential dimensions. More precisely, we will consider the following two specific dimensions: the *physical* and the so-called

---

[1] In fact, Greek philosophers like Democritus were not only interested in explaining physical substances, but extraphysical ones as well.

extraphysical per se (Vieira, 2002). As it is well known, these two dimensions appear to interact very weakly, seeing that it is not at all easy for an extraphysical consciousness to objectively directly manifest in the physical dimension (the converse being equally true, of course). A natural question then arises:

Why can't an extraphysical substance easily interact with a physical substance, and vice versa?

The question, at first sight, may appear puzzling, because the physical and extraphysical dimensions both contain, at least in principle, unlimited amounts of energy. Therefore, the observed weak interactivity cannot be explained by a mere argument of energy-shortage of one dimension compared to the other, not to mention that the problem manifests in both directions: from the extraphysical to the physical but from the physical to the extraphysical as well. Thus, a more refined version of the above question could be:

Why is the energy transfer from the extraphysical to the intraphysical dimension, and vice versa, in general, so inefficient?

2 The frequency model

To answer the above question, one usually invokes the concept of *frequency*. Let us briefly describe how the typical heuristic goes. One starts by assuming that all entities within reality possess vibrational properties,[2] expressible in terms of a set of characteristics, natural resonance *frequencies*, forming the frequency *spectrum* of the entity. To be more specific, let us denote by σ_A the spectrum of a given entity A. This means that A can *vibrate* only at frequencies belonging to the set σ_A. In the same way, consider another entity B, possessing a spectrum σ_B. Now, since A and B can only vibrate at frequencies within their own spectrum, they can interact together, i.e., they can efficiently exchange energy, if and only if the intersection $\sigma_A \cap \sigma_B$ of their spectra is not the empty set (see Figure 1).

Then, one can explain the inefficiency of the energy transfer

---

[2] This hypothesis is also known as the *principle of vibration*, in hermetic philosophy.

270

between the physical and the extraphysical dimensions by hypothesizing that the vibrational frequencies characteristic of an extraphysical entity are, generally speaking, much higher than those of a physical one, so that their spectra do not overlap. In other words, physical and extraphysical entities cannot efficiently exchange energy as they do not share a common frequency channel through which they could communicate (see Figure 2).

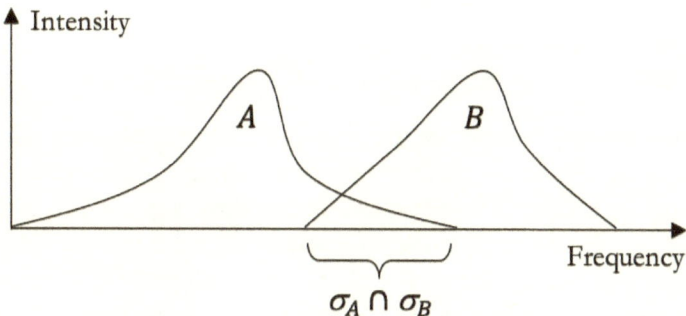

Figure 1 A schematic illustration of two entities whose spectra overlap (i.e., their intersection is not empty).

The hypothesis that entities belonging to the extraphysical dimensions do vibrate at higher frequencies in comparison to physical entities is supported by a certain number of paraperceptions, as reported for instance by lucid (out of body) projectors. Let us mention, as a typical example, the sensation of an intense, increasing and continuous vibration (vibrational state) which can be experienced during the period of exteriorization of the psychosoma.

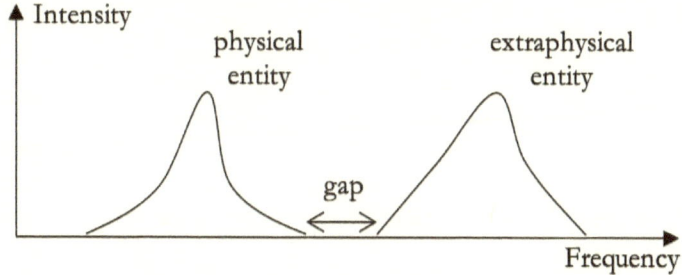

Figure 2 A schematic illustration of a physical and an extraphysical entities, whose spectra do not overlap.

But although it is undeniable that the physical and extraphysical dimensions cannot easily affect one another, it is also incontestable that there exist multidimensional structures for which the interdimensional energy transfer seems to work with great efficiency. Consider for instance our soma, whose existence heavily depends on being sustained by the psychosoma. Despite the hypothesized frequency gap between these two vehicles, an intense flux of informed energy appears to be efficiently and continuously maintained between them, during the entire intraphysical life of the consciousness.

Why does the energy transfer between the psychosoma and the soma, and vice versa, arise with great efficiency within our holosomatic structure?

The well-known answer to the above question is to point out that between the psychosoma and the soma there is an intermediary energizing agent, the *holochakra* (also called the *energosoma*), and that it is precisely thanks to its role of *mediator* that the two vehicles can efficiently exchange energy. Within the paradigm of the frequency model, one can explain the functioning of the holochakra by simply assuming that the quasiphysical (or quasiextraphysical) substances it is made of possess a frequency spectrum which is intermediary with respect to the somatic and psychosomatic ones, so that it has a non-empty intersection with both of them. In other words, the holochakra would act as a bridge between the somatic and psychosomatic vehicles, by filling in the frequency gap between these two entities (see Figure 3).

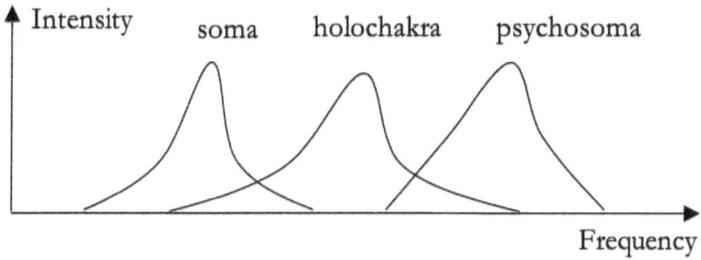

Figure 3 The holochakra playing the role of a frequency-bridge between the soma and the psychosoma.

AutoRicerca, Issue 20, Year 2020, Pages 267-287

3 The classical density model

The frequency (resonance) model that we have briefly described is well known and has been widely discussed in the literature. Almost all works investigating "subtle" energies do mention, in a way or another, the idea of frequencies and resonances. Just to give an example, see Vieira (2002), p. 205 and pp. 979-987. Of course, determining which type of fields these frequencies would refer to is an issue that the frequency heuristic model does not address. Obviously, these are not frequencies related to known physical fields, such as the electromagnetic one. On the other hand, a heuristic model is not meant to necessarily offer a tested solution to a problem, but to provide a way of thinking that can facilitate the access to more advanced knowledge, in the ambit of more elaborated and mature scientific theories.

The purpose of the present work is to present a heuristic model which constitutes an alternative to the frequency model, providing a conceptual framework that can also explain the observed inefficiency of the interdimensional energy transfer. The model is extremely simple: it consists in saying that the most important difference between matter and paramatter resides in their different *densities*. More precisely, our basic assumption is that, generally speaking, *matter is much denser than paramatter*. However, we do not mean by this that extraphysical substances would be more *rarefied* than physical ones, as it would be the case, for instance, for a gas in comparison to a liquid or to a solid. What we intend is that a typical physical particle (like an electron) is, by many orders of magnitude, much more massive than a typical extraphysical particle (for instance, a paraelectron, assuming it would exist). Therefore, given a physical and an extraphysical substance, each having the same number of particles per unit volume, what we are here assuming is that the density of the former is much higher than the density of the latter, because the *inertial masses* of physical particles are, in general, much higher than those of extraphysical ones.

As for the frequency hypothesis, the density hypothesis is

supported by a number of paraperceptions, as reported by lucid projectors (Vieira, 2002). Let us mention, as a typical example, the phenomenon of *extraphysical bradykinesis*, a condition of slowness perceived by the consciousness while moving in the projected psychosoma. The cause of this slow-motion effect is usually perceived as being related to a higher density of the *extraphysical sphere of energy*, in comparison to the lightness of the substance forming the moving psychosoma.

Let us now assume, to simplify our model as much as possible, that both the material and paramaterial substances are made of *classical point-like particles*. According to our hypothesis, the only relevant difference between the two substances resides in the mass of their constituents. More precisely, let us denote by m the typical mass of a *paramaterial* particle and by M the typical mass of a *material* one. Our hypothesis is that the mass ratio m/M is very close to zero.

To determine the efficiency of the energy transfer between paramatter and matter, we need to analyze what happens during a *collision* between a paramaterial particle and a material one, and ask:

How much energy can the paraphysical particle of mass m transfer to the physical particle of mass M?

To answer this question, let v be the velocity of the paraphysical particle moving toward the physical one, and let us assume that the latter is initially at rest (see Figure 4).

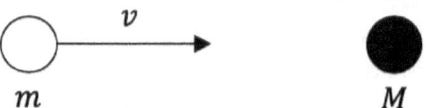

Figure 4 An extraphysical particle impinging with velocity v on a physical particle, initially at rest.

After the collision (here assumed one-dimensional and purely elastic, for simplicity), the paraphysical particle will move in the opposite direction with a smaller velocity $v' < v$, whereas the physical particle, which was initially at rest, will acquire a non-zero velocity v'' (see Figure 5).

Figure 5 Following the collision, the lighter extraphysical particle bounces back, after having put the physical particle into movement.

The initial energy E of the incoming paraphysical particle is given by its kinetic contribution $E = \frac{1}{2} mv^2$. Similarly, the final energy E'' acquired by the physical particle is given by $E'' = \frac{1}{2} mv''^2$. We are interested in calculating the *energy efficiency* η of the collision process. More precisely, we want to determine the ratio $\eta = E''/E$, given by the *output energy* E'' of the physical particle divided by the *input energy* E of the paraphysical one. By definition, the dimensionless parameter η is a number between 0 and 1. The case $\eta = 0$ corresponds to a zero-energy transfer, whereas the case $\eta = 1$ corresponds to a total energy transfer.

To calculate η, one needs to exploit two important principles: *energy conservation* and *momentum conservation*. After some algebra, one easily finds for η the following simple formula:

$$\eta = \frac{4\lambda}{(1 + \lambda)^2}$$

where we have defined $\lambda = m/M$. We can observe that the efficiency only depends on the mass ratio λ, and that its maximum value $\eta = 1$ is reached when $\lambda = 1$, that is, when the collision is between two particles of same mass (think for instance to the well-known *Newton's pendulum*).

In the situation of our concern, however, which is the interaction between paramatter and matter, the mass ratio λ is typically close to zero, as by hypothesis M exceeds m by many orders of magnitude. Now, if λ tends to zero, it immediately follows from the above formula that the efficiency η of the energy transfer also tends to zero.

As an example, let us hypothesize that the mass of a typical paraphysical particle is, on average, *one thousandth* of the mass of a physical one, based on the speculation that the average weight of a projected intraphysical consciousness' psychosoma appears to be, approximately, one thousandth of the weight of the human body that houses it [see (Vieira, 2002), page 288]. Then, replacing $\lambda = 0.001$ into the previous formula, one obtains for the efficiency, the approximate value: $\eta \cong 0.004 = 1/250$. This means that to transfer 1 unit of energy to a physical particle, an extraphysical particle would need to carry at least 250 units of energy, that is, 250 times more.

Thanks to this elementary model of classical colliding particles, we can already understand why the energy transfer between paramatter and matter is so difficult. Due to the hypothesized great mass difference between the physical and extraphysical energetic carriers, the efficiency of the process is very low, and one needs huge amounts of energy to produce even the tiniest effect.[3] So, as we did for the frequency model, we can now ask the following question:

Can we use our simple model of point-like classical particles to gain some insight into the functioning of the holochakra, i.e., in its role of energetic mediator between the soma and the psychosoma?

To this end, let us assume that the energetic substance composing the holochakra is made of quasiphysical particles having an intermediary mass compared to those composing the soma and psychosoma. As we shall see, this assumption is sufficient to explain the observed gain in efficiency for the energy transfer due to the mediation of the holochakra. More precisely, let us suppose that the energy transfer from the incident paraphysical particle of mass m to the target physical particle of mass M takes place with the help of another particle of mass μ, located between them (see Figure 6). In other terms, the incident paraphysical particle of energy E first hits the mediator particle (supposed at rest), which next hits the final physical target particle (also supposed at rest).

---

[3] It is worth emphasizing that the obtained formula for the energy efficiency η also remains valid if the incoming particle of energy E is the physical instead of the paraphysical one, so that exactly the same inefficiency in the energy transfer holds when going from the physical to the extraphysical, as one would expect. One can in particular observe that

$$\eta = \frac{4\lambda}{(1 + \lambda)^2} = \frac{4\lambda^{-1}}{(1 + \lambda^{-1})^2}$$

which means that the roles of m and M are interchangeable in the formula.

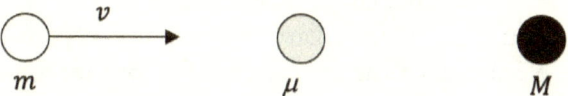

Figure 6 A quasiphysical particle is placed between the incoming extraphysical particle and the target physical particle.

The efficiency η_2 of the entire process is now given by the product of the efficiencies of the two successive collisions. Thus:

$$\eta_2 = \frac{4\alpha}{(1+\alpha)^2}\frac{4\beta}{(1+\beta)^2}$$

where $\alpha = m/\mu$ and $\beta = \mu/M$. If the three masses are equal, then $\alpha = \beta = 1$ and $\eta_2 = 1$, i.e., the energy transfer is maximum. But in the situation of our interest, all three masses are different, and the energy transfer is not equal to unity. However, we may ask for which value of the mediator mass μ the efficiency η_2 reaches its maximum value. After a straightforward calculation, one finds that the maximum is reached when the mediator mass μ is exactly the *geometric mean* of the masses m and M, i.e., $\mu = \sqrt{mM}$. Then, we have $\alpha = \beta = \sqrt{\lambda}$, and replacing these values into the previous formula, one finds for the efficiency:

$$\eta_2 = \frac{16\lambda}{(1+\sqrt{\lambda})^4}$$

We can compare this expression with the one previously derived for the efficiency η in the absence of the mediator particle. As before, let us consider the case where the mass m of the paraphysical particle is one thousandth of the mass M of the physical one. Replacing the value $\lambda = 0.001$ into the above expression, one finds for the efficiency: $\eta_2 \cong 0.014 \cong 1/71$. Thus, we obtain that to transfer 1 unit of energy to a physical particle, an extraphysical particle using a single optimal mediator only needs to carry 71 units of energy, instead of 250. In other terms, thanks to the mediator, the efficiency of the energy transfer has been increased by 350%. And in fact, it can be shown that it can be increased up to 400%; see (Bashkansky et al, 2007).

The above simple calculation shows that by using a mediator particle with a suitable intermediary mass, one can considerably increase the efficiency of the energy transfer. But then, we may further ask:

Can we further increase the efficiency of the process by increasing the number of mediators?

To answer this question, let us assume that between the incoming paraphysical particle of mass m and the final target particle of mass M, there is an entire linear arrangement of $n - 1$ quasiphysical intermediary particles of variable mass (see Figure 7).

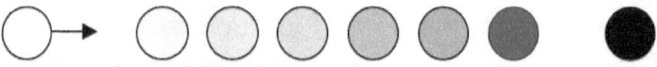

Figure 7 Quasiphysical particles of increasing mass are placed between the incoming extraphysical particle and the target physical particle.

We can choose the masses of the intermediary particles in the following way. Let $\mu(x)$ be a well-behaved function, defined in the closed interval $[0,1]$, such that $\mu(0) = m$ and $\mu(1) = M$. Without losing generality, we can define the masses of the $n + 1$ particles (the incoming paraphysical particle plus the $n - 1$ quasiphysical mediator particles plus the target physical particle) as follows:

$$m_k = \mu\left(\frac{k}{n}\right), k = 0,1,2,\dots,n$$

To calculate the energy efficiency η_n of the multiple process, we need to observe that it is simply given by the product of the efficiencies of the n sequential collisions:

$$\eta_n = \eta_{0,1}\,\eta_{1,2} \cdots \eta_{n-1,n}$$

where $\eta_{k,k+1}$ is the ratio of the energy transferred to the particle of mass m_{k+1} to the energy of the incoming particle of mass m_k, which is given by:

$$\eta_{k,k+1} = \frac{4\alpha_{k,k+1}}{(1 + \alpha_{k,k+1})^2}$$

with $\alpha_{k,k+1} = m_{k+1}/m_k$. It is then an easy matter to show that, as the number n of mediators tends to infinity (i.e., as $n \to \infty$), the efficiency η_n tends to 1 (i.e., it becomes maximal), provided $\mu(x)$ is a differentiable function (Sassoli de Bianchi, 2007).

To perform an explicit calculation, let us choose the special case where $\mu(x) = \lambda^{-x} m$. It is then straightforward to obtain the following formula for the efficiency:

$$\eta_n = \left(\frac{2\sqrt{\lambda^{\frac{1}{n}}}}{1 + \lambda^{\frac{1}{n}}} \right)^{2n}$$

For the values $n = 1$ and $n = 2$, we recover the two previously derived expressions for η and η_2, respectively. But, as the number of intermediary quasiphysical particles increases, i.e., as n tends to infinity, then $\lambda^{\frac{1}{n}}$ tends to $\lambda^0 = 1$ and η_n also tends to 1, in accordance with the previously mentioned general result.

Consider once more the case $\lambda = 0.001$. Then, one can use the above expression to calculate the following values for η_n:

$\eta_1 \cong 0.004$, $\eta_2 \cong 0.014$, $\eta_5 \cong 0.109$, $\eta_{10} \cong 0.310$,
$\eta_{50} \cong 0.788$, $\eta_{100} \cong 0.888$, $\eta_{200} \cong 0.942$, $\eta_{400} \cong 0.971$

Thus, for an array of approximately 100 mediators, one finds that the efficiency of the energy transfer is already very close to its maximum value.

Let us briefly summarize our findings so far. The hypothesis at the basis of our simple model is that substances pertaining to different dimensions exhibit different densities, hence different inertial masses of its constituents, a physical particle being, on average, much more massive than an extraphysical one. This hypothesis, together with the energy and momentum conservation laws (which are here assumed to apply also in the extraphysical dimensions), can explain the observed weak interactivity among matter and paramatter. According to this model, the holochakra could be understood as an interdimensional bridge made of multidimensional substances of variable density.

As we said already, the density gradient of the holochakra should however not be understood as the result of a rarefaction of its constituents, but instead as a variation of their intrinsic inertial properties. To assure a maximum efficiency in the downloading and uploading of energies, it is sufficient for the mass of the particles composing the holochakra (also called *silver cord*) to vary smoothly, when going from the physical to the extraphysical, and vice versa (see Figure 8).

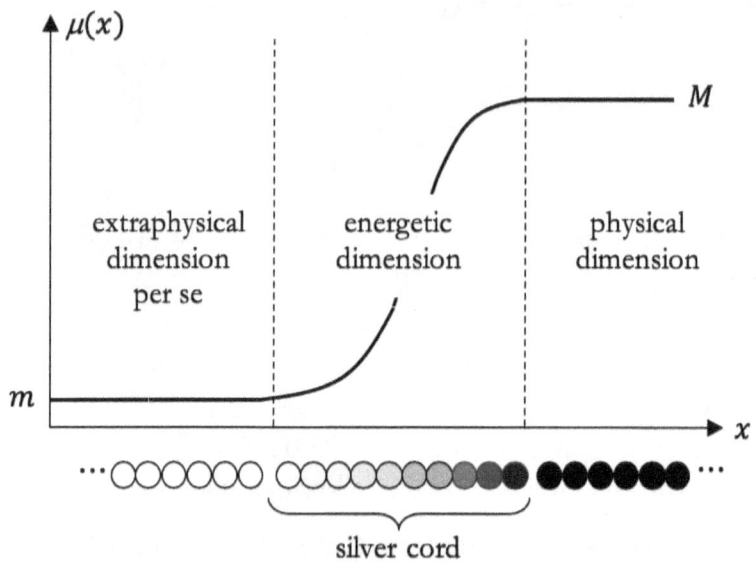

Figure 8 The holochakra (silver cord) as a variable density/mass structure bridging the physical and extraphysical dimensions.

5 The quantum mass model

The mass model we have presented is very simple and its interest resides mainly in its heuristic content. But apart from the oversimplification of having considered one-dimensional, non-quantum, non-relativistic particles, it is also legitimate to ask on which basis one can assume that the inertial mass of a particle globally decreases when going from a lower to a higher existential dimension. In other terms:

What kind of picture can we adopt to justify the hypothesized dimensional-dependence of a particle's inertial mass?

280

An interesting answer comes from the study of non-homogeneous crystals and semiconductor heterostructures. Indeed, in the study of the transport properties of quantum particles (like for instance electrons) propagating in such systems, one can usually take into account the interaction of the particle with its host structure in terms of an *effective mass*. In other words, according to this approximation, everything happens as if the particle moving inside the structure acquires a different effective inertial mass. Accordingly, it also follows that when the media inside which a quantum particle moves is non-homogeneous, then its effective mass is no more a constant of the motion, but a function of its position [see for instance Lévy-Leblond (1995) and the references cited therein].

Adopting such a conceptual framework, one can hypothesize by analogy that the overall structure characterizing an entire reality dimension (like for instance the physical or the extraphysical *per se*) is similar to a huge ordered crystalline structure inside which the different energetic entities can manifest and move. This would mean that entities can experience different effective masses according to the specific, crystal-like, dimensional structure in which they are immersed, and this could support our hypothesis of a variation of particles' inertial properties when traveling in different existential dimensions.

The above discussion allows us to propose an additional quantum model, suggesting an extra mechanism for the observed inefficiency of the interdimensional energy transfer. Instead of considering classical colliding particles, we can now consider the propagation of a flux of independent quantum particles with *position-dependent mass*. When inside the extraphysical domain, the particles possess an effective mass m, but when inside the physical domain, as the structure is different, they acquire a greater effective mass M.

In quantum mechanics, the equation describing the motion of a (here one-dimensional) "free" particle of energy E with position-dependent mass $\mu(x)$, is given by a modified version of the stationary Schrödinger equation (Lévy-Leblond, 1995):

$$-\frac{1}{2}\partial_x\left[\frac{1}{\mu(x)}\right]\partial_x\psi_E(x) = E\psi_E(x)$$

where $\psi_E(x)$ denotes the particle's wave function. Let us consider the situation where an extraphysical particle tries to penetrate into the physical dimension, without passing through a mediator structure like the holochakra. In this case, the particle experiences an abrupt variation of its effective mass as a consequence of the sharp (discontinuous) interdimensional interface. This means that the effective position-dependent mass $\mu(x)$ of the particle is described by a step-like function (see Figure 9).

As before, we are interested in determining the efficiency η of the energetic transfer, which is now given by the ratio of the intensity of the incoming particles' flux to the intensity of the transmitted flux. This ratio generally differs from unity, because not all particles composing the incoming flux are transmitted through the interdimensional interface. Part of the incoming flux is indeed reflected back. Let us emphasize that the reflection mechanism is not here the consequence of the interaction of the incoming particles with some sort of force field. The particles, indeed, are supposed to move freely, and their reflection at the interdimensional boundary is just the result of a genuine quantum effect, due to the discontinuous variation of their effective mass.

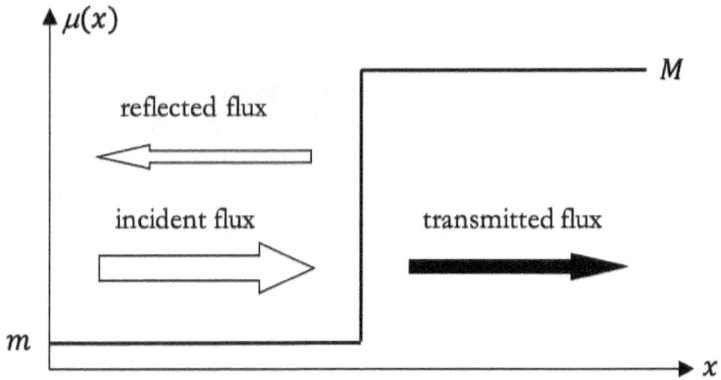

Figure 9 Because of the mass discontinuity, the incident flux of extraphysical particles is split into reflected and transmitted components.

More precisely, the efficiency η is given by the *probability* of a quantum particle of energy E being transmitted through the mass step-barrier. Using the above modified Schrödinger equation, it is not difficult to calculate such a probability (Lévy-Leblond, 1992), which is given by:

AutoRicerca, Issue 20, Year 2020, Pages 267-287

$$\eta = \frac{4\sqrt{\lambda}}{\left(1 + \sqrt{\lambda}\right)^2}$$

Where, as before, $\lambda = m/M$. We emphasize that exactly the same formula holds for particles traveling in the opposite direction, that is, from the physical to the extraphysical dimension.

Now, for $\lambda = 1$, as expected, $\eta = 1$. Furthermore, as the mass ratio λ tends to zero, the efficiency η also tends to zero, which means that in this limit all particles are reflected back. For the specific ratio $\lambda = 0.001$, one finds $\eta \cong 0.12 \cong 1/8$, which is approximately **30** times better than what we have calculated in our previous classical model. However, we should not compare these two models, neither quantitatively nor qualitatively, as their logics are very different.[4]

Again, we can wonder how the holochakra would function within our position-dependent mass model. To increase the efficiency of the transmission mechanism, one can think of the holochakra as a non-homogeneous crystal-like morphothosene, responsible for a smooth and gradual variation of particles' inertial mass, as a function of their interdimensional position. Indeed, using an adaptation of the so-called *WKB* semiclassical approximation, one can show that for a sufficiently smooth and slowly varying mass function $\mu(x)$, the totality of the incident flux is transmitted, so that the energy efficiency of the process becomes maximal.

12 Concluding remarks

Why do objects have a mass? What is the typical answer physicists today give to this basic and at the same time very difficult question?

---

[4] In the previous classical model, the energy transfer mechanism was the consequence of a two-particle collision process, whereas in the present quantum model it is the consequence of a single-particle transmission process through a mass-barrier. One can of course imagine combining these two models into a more sophisticated and integrated picture, considering for instance the more general situation of an N-body system of quantum scattering particles with position-dependent masses. The study of such a model, however, is far beyond the scope of the present paper.

According to physicist and philosopher *Ernst Mach* (1838-1916), inertia cannot exist in an empty space, as it results from the mutual gravitational interaction between all entities populating the universe. This is the so-called *Mach principle*. In 1961, Mach principle has been successfully integrated into Einstein general relativity equations by *Carls Brans* and *Robert Dicke* (Brans and Dicke, 1961) in the form of a variable (in space and time) field determining the intensity of the gravitational forces and consequently (because of the equivalence principle) the inertial masses of the different material objects. This is the so-called *Brans-Dicke's field*.

In the same years, but in a completely different context, *Peter Higgs* (Higgs, 1964) discussed how a field permeating the entire universe (previously introduced by *Jeffrey Goldstone* as a special solution to certain field equations) could be responsible, through its interaction with all kind of particles, for a mechanism of mass generation (a symmetry breaking phenomenon known as the *Higgs mechanism*).

Later on, at the end of the seventies, and thanks to the work of a generation of physicists well-formed both in particle physics and cosmology (in particular *Anthony Zee, Lee Smolin, Alan Guth, Andreï Linde* and *Gabriele Veneziano*), it was realized that the Brans-Dicke's field and the Goldstone-Higgs field are just two different descriptions of a same phenomenon, possibly explaining the origin of inertia in our universe. In addition to the field of Brans-Dicke or Goldstone-Higgs, many authors have proposed alternative mechanisms to explain inertia. Let us mention, as an example, the zero-point-field model of *Haisch, Rueda* and *Puthoff* (Haisch et al, 1994).

Anyway, our intention here was not to review today's leading-edge theories regarding this difficult problem, but just to emphasize that according to the most advanced models it seems natural to assume the existence of a field varying in space and time, filling our entire reality, responsible for the attribution of the observed inertial properties to the different entities. This is in good accordance with the heuristic at the base of the present work, considering that the effective mass of physical and extraphysical particles would be the consequence of their overall interaction with a variable and multidimensional field, shaping and demarcating the different physical and extraphysical dimensions.

We already know from relativity theory that the mass of an entity is not a conserved quantity. Indeed, according to Einstein's most famous equation, $E = mc^2$, mass and energy are completely equivalent to one another. In our model, however, when we refer to the mass of a particle what we mean is its *rest mass*, not its relativistic mass. In other terms, the mass variation we have hypothesized is not to be confused with the relativistic increase of the mass of a particle as a function of its velocity.

We can observe that our mass model assumes a decrease of the effective mass of an entity when going from the physical to the extraphysical dimensions. The frequency model, on the other hand, assumes an increase of the frequency of vibration of an entity when going from the physical to the extraphysical. So, it is natural to ask:

Are these two assumptions compatible?

To answer this question, we can consider the paradigmatic example of a *spring-mass system* of elastic constant k and mass m. The frequency f of its harmonic oscillations is given by the formula:

$$f = \frac{1}{2\pi} \sqrt{\frac{k}{m}}$$

Therefore, as the mass of the system decreases, its frequency of oscillation increases, and vice versa. This shows that if we describe an entity as a system possessing a certain amount of potential energy, which can be fully converted into an internal oscillatory movement, then the description of the mass and frequency models are compatible.

The mass (or density) model allows us to understand the energetic separation between dimensions in terms of an inefficiency of the energy transfer. There are of course many examples of parapsychic phenomena revealing the extremely low efficiency of the energy transfer mechanism. One can cite the example of telekinesis, where the rate of success is notoriously very low, and enormous expenditure of consciential energies are usually required to move even the smallest and lightest physical object.

The mass model also points out the necessity of having structures like the holochakra: mediators of a multidimensional nature whose mass (density) varies smoothly and gradually, in order to connect the physical and extraphysical dimensions by improving the efficiency of the interdimensional energy transfer.

Apart from the internal structure of our holosoma, one can of course identify many other situations where the presence of a mediator structure allows for an improvement of the interdimensional communication. A typical example is the so-called *penta* technique [see Vieira (2002), p. 594, and Figure 293], a process during which a helper (an entity pertaining to the extraphysical per se dimension) transmits healing consciential energies to an ill consciousness (an intraphysical projected consciousnesses or an extraphysical consciousnesses having not yet undergone the so-called second desoma). To succeed in the transfer of energy, the helper uses the mediation of the penta practitioner, whose holochakra provides the intermediary-density connecting-bridge between the "light-weight" helper and the "heavy-weight" ill consciousness.

Even more interesting is the *energization by three* technique [see Vieira (2002), p. 696, and Figure 357], where the helper uses two mediators at the same time: a "subtler" projected intraphysical consciousness and a "denser" non-projected intraphysical consciousness. According to our simplified model, it is natural to conjecture that such a double-mediator configuration permits a further gain in efficiency, in comparison to the standard, single-mediator, penta technique.

As a last example of a quasiphysical mediator structure, let us mention the *assistential bioenergetic field*, as implemented for instance during IAC's *Consciousness Development Course–Advanced 2: Assistantial Energetic Field*. Thanks to the connection established between a team of advanced extraphysical helpers (possibly also employing some kind of paratechnology) and the holochakra of the intraphysical epicenter, a temporary multidimensional energetic bubble is produced around the epicenter. We can conjecture that the bionergetic field can reduce the gap between the extraphysical and physical dimensions because it would be made of substances of varying density, i.e. because it would possess a specific multi-layered structure.

To conclude, let us emphasize once more that the validity of our heuristic mass model is based on a number of speculative assumptions (this is the case also for the frequency-model). Not only we have assumed that energy and momentum are conserved quantities in the extraphysical dimensions, but also that physical concepts like mass and density (or frequency and intensity in the frequency model) are still meaningful in non-physical domains. Of course, nothing is less sure than this, considering our limited knowledge of paraphysics.

Acknowledgments

The author is grateful to *Nelson Abreu* for useful discussions and for his critical reading of the manuscript.

Bibliography

Bashkansky, E. and Netzer, N. (2006). The role of mediation in collisions and related analogs, American Journal of Physics 74, pp. 1083-1087.

Brans, C. and Dicke, R.H. (1961). Mach's Principle and a Relativistic Theory of Gravitation, Physical Review 124, pp. 925-935.

Haisch, B., Rueda, A. & Puthoff, H.E. (1994). Inertia as a zero-point-field Lorentz force, Physical Review A 49, pp. 678-694.

Higgs, Peter W. (1964). Broken Symmetries and the Masses of Gauge Bosons, Physical Review Letters 13, pp. 508-509.

Lévy-Leblond, J.-M. (1992). Elementary quantum models with position-dependent mass, European Journal of Physics 13, pp. 215-218.

Lévy-Leblond, J.-M. (1995). Position-dependent effective mass and Galilean invariance, Physical Review A 52, p. 1845.

Sassoli de Bianchi, M. (2007). Comment on 'The role of mediation in collisions and related analogs, by E. Bashkansky and N. Netzer," American Journal of Physics 75, p. 1166.

Vernon Vugman, N. (1999). Conscientiology and Physics: A Desirable Couple? Journal of Consciousness, Volume 1, No. 4, p. 289.

Vieira, W. (2002). *Projectiology, A Panorama of Experiences of the Consciousness outside the Human Body*, Rio de Janeiro, RJ – Brazil, International Institute of Projectiology and Conscientiology.

Note: This article was previously published in: Journal of Consciousness 11, 2009, pp. 297-316, with the title: "Interdimensional energy transfer: a simple mass model." An Italian translation is also available in: *Lo Stato Vibrazionale*, AutoRicerca, Numero 6, Anno 2013, pp. 89-112.

Invitation to read

Vibrational State and Energy Resonance

Self-tuning to a higher level of consciousness. A practical and theoretical guide to mastering and understanding the human energy body

Author: Nanci Trivellato; *pages:* 494; *year:* 2017; *publisher:* International Academy of Consciousness; *ISBN:* 9781544944036

Reviewed by: Massimiliano Sassoli de Bianchi

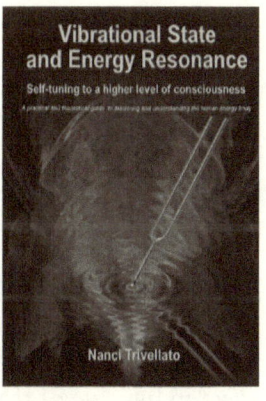

When speaking of the Vibrational State, we are not referring to a condition of our physical body, but of our energosoma: a complex structure bridging our intraphysical condition with the "subtler," extra-ordinary (extra-physical) aspects of our manifestation.

The first time I experienced a vibrational state, I wasn't aware of the existence of this particular condition. It happened to me more than fifteen years ago, in a totally spontaneous way, with an intensity I have not been able to experience again since. It was during a group therapy session, and I remember that I was doing a visualization exercise, eyes closed, which consisted of visualizing a person from my past, to check if a residual "energetic chain" was still binding us. While doing the exercise, I did not interact with the other persons in the room, and I was very calm. I visualized the chain, took it in my hands, then, in a decisive way, I severed it. Following this simple movement, something totally unexpected happened to me, which at the time I described with the following exact words:

"All of a sudden my body started to vibrate with an unbelievable energy, and successively a deep pain and sadness escaped from me. The energy circulating in my body and throughout my hands was so intense that I could not stand, and I fell on my knees. I was grateful for this gift of energy, which allowed me to free myself from my pain and sadness. I observed that my body was crying heavily, but within myself I felt totally supported, and happy that I could offer myself an experience so intense and beautiful. An amazing vibrational form of energy was crossing my entire body. I didn't know how it looked from the outside, but my impression was that I was about to dematerialize, and asked myself if the other persons in the room were still able to see my body. Later on, the therapist told me that he had felt a very strong energy around my person, which at first even caused him to recoil. The phenomenon lasted about a quarter of an hour, with the vibrations gently diminishing, leaving me a bit dazed and without any hunger that evening."

Five years following this experience, I came across some intriguing "subtle energy" research, based on experiments with inert gases. The person involved in these studies, with whom I corresponded for some time, described to me for the first time, in specific terms, that "sense of vibration" which is usually associated with the "take off" phase of an out-of-body experience. He also told me that a "vibrational state exercise" existed, that one could practice to promote such a condition. A few years later, this brought me to London, at the office of the *International Academy of Consciousness* (IAC), to attend their award-winning course Consciousness Development Program (CDP). It was then that I met for the first time *Nanci Trivellato*, in her role of teacher and director of the research department.

Attending the CDP course helped me to clarify many aspects of my previous energetic experiences. In particular, it was now clear to me that the powerful vibration that shook my entire person years before was almost certainly the result of an intense energetic shower promoted by the extraphysical helpers who were working with the group, which activated in my energosoma and psychosoma an intense and long lasting *vibrational state* capable of deeply cleaning my energetic sphere of old, congested energies.

If the self-researcher in me was happy to have found, in the author and her colleagues, a group of very competent and committed people, teaching and researching the evolution of consciousness in a very open and critical way (a rare binomial in the present "new age"), always balancing the theoretical and practical aspects of their activities, the physicist in me, inevitably, remained with more questions than answers, regarding the true nature of many of the subtle phenomena that I could experience firsthand.

For instance: What was truly vibrating in my vibrational state? And what did the frequency of these vibrations refer to? Also, when we perceive an increase of the frequency of the vibrations, when "moving" from the intraphysical to the extraphysical state, is this variation a subjective perceptual phenomenon, or an objective process, associated with quantifiable variations in the relative speed of some non-ordinary substance circulating in our energosoma? Can we still define a frequency, as we do in physics, when dealing with non-ordinary entities like our "subtle" vehicles, which so far have evaded any attempt to be measured by physical apparatuses, or to be consistently modeled in theoretical terms? And considering that the flow of time appears to constantly fluctuate in an out-of-body situation, with projectors having reported extremely long extraphysical experiences happening in very short intraphysical times, and vice versa, how can we even attach a proper sense to a notion of frequency and vibration, in such a hybrid interdimensional context?

The above are just examples of the questions I was asking myself at the time, when trying to understand the nature of the phenomena that we consciousnesses can experience, like the vibrational state, but cannot easily explain, particularly if we adopt an excessively reductive, mono-materialistic perspective. With pleasure, I observed that Nanci Trivellato, and her colleagues at the IAC, were not insensitive to these interrogatives, as they encouraged me to use my scientific training, as a physicist, and my mental openness, as a self-researcher, to possibly find in the future some new elements of clarification. This is important to emphasize, as nowadays we can find on this planet several organizations proposing different approaches to spirituality, but in most them the aspect of the research remains largely undervalued. This was not the case of Trivellato and her colleagues, who instead were strongly committed in furthering their theoretical and practical understanding of the multifaceted subject they were teaching.

The difference between a person who is truly committed to a path of research (inner and outer), and one who is only mimicking it, is easy to discern, if one has the chance of observing the evolution of the person throughout the years. I have followed many courses and workshops offered by the author in the last ten years, and what I could observe is a steady progression not only concerning the quality of her teaching, but also of her understanding and contextualization of the subjects she teaches. I remember for instance how impressed I was when, during the *2nd International Symposium on Conscientiological Research*, in 2008, I listened to her detailed analysis of the "vibrational state exercise," that she more technically renamed *voluntary energetic longitudinal oscillation* (VELO). The difference between what I had been taught until then, and could personally understand about this fascinating technique, and what the author was now able to explain and illustrate, in a very systematic and precise way, was truly remarkable, and demonstrated to me the amount and the quality of the research (and self-research) she has been conducting up to that time. Not for nothing, the article in which these findings were subsequently published, entitled "Measurable Attributes of the Vibrational State Technique" (republished in the present volume) received the first prize of the *2nd LAC Global Award for Scientific Contribution to Consciousness Science*.

At that time, I was offering a few courses in my small laboratory, which also included the practice of techniques taken from the tradition of yoga, probably one of the oldest self-research traditions of our planet. It was interesting for me to observe that the VELO technique, so lucidly analyzed and explained by Trivellato, had some points of correspondence with certain yoga practices, usually referred to as *kriya*, *pranayama* and *pranavidya*. Thus, I decided to also include her analysis of the energosomatic attributes in my courses, exploring the possibility of integrating the VELO technique with some of the yoga's pranayama, in the spirit of an ongoing research possibly bridging the knowledge of the past with our most recent understandings.

AutoRicerca, Issue 20, Year 2020, Pages 289-294

Shortly after its publication in the *Journal of Consciousness*, in English and Portuguese, the author's award-winning article was also published in Italian, in the first issue of the journal *AutoRicerca*, of which I am the editor, as it was very clear to me that this was a text of great importance in the "inner technology" panorama. Also, her demystified approach to the energy work was more than needed in a country like Italy (but not only), where, because of religious conditionings, many individuals still believe that "moving energy" can be dangerous for one's "spiritual health."

But let me spend a few more words on this little gem of a book. This is a text with many qualities. The most important one, from my viewpoint, is that it is a truly original contribution. Too often we find, on the shelves of the libraries, volumes that are new only in their titles and dates of publication, but whose contents are nothing more than a long paraphrase of previous writings. This is particularly true in the field of spiritual research, where hardly anything new is ever written, partly because there are many who like to speak and write, but very few who are dedicated to an authentic exploration of the frontiers of our knowledge. The author is one of these few, and *Vibrational State and Energy Resonance* is the proof that new ideas and real progress are possible even in this difficult field of investigation.

Trivellato's book is also a much-needed contribution, as it is a rare example of what a seriously conducted co-operative research, using first-person and second-person methods, can accomplish. This is important because most academic researchers today remain pretty much adverse to any form of special training to obtain more reliable, lucid and self-controlled perceptions, particularly when these concern the more "subtle" aspects of our manifestation. But it is also clear that a scientific debate can only be based, in ultimate analysis, on experimental evidence, and since for the time being we humans appear to be the only instruments that can measure "subtle" energies and work with them in a controlled way, we have to learn, in fits and starts, to have finer grained and better controlled perceptions if we want to study their phenomenology in a more systematic way, and possibly find new explanatory models.

In that respect, Trivellato's book is not only a text filled with new intuitions and findings, obtained thanks to her long experience as a self-researcher and facilitator of other energy workers: it is also a very detailed practical manual, in which one can find clear indications on how to efficaciously work with one's bioenergy and master the fundamental VELO technique, to obtain vibrational states of increased quality.

No need to say, working with the energosoma (energy body), and studying its properties, is a very delicate task. Indeed, this is an entity that, in our ordinary intraphysical state, is deeply entangled (i.e., deeply connected) with our physical body. Therefore, our perceptions, particularly in the beginning of our practice, will necessarily be of a hybrid nature, coming from both our energosoma and our soma. This is one of the obstacles a self-experimenter needs to face, which is almost like a paradox to solve: that of discriminating two entities which we initially perceive as one. But as Trivellato demonstrates in her book, with some perseverance, and the right information, it is certainly possible to experientially separate them, and concentrate with increased efficiency and efficaciousness only on the subtler energosomatic paraperceptions, and the energopsychomotor abilities that are associated with them.

This is precisely what the author has done for us in her brilliant work: thanks to her didactical and thorough analysis, she has disentangled for us the "holosomatic machine," showing how we can work, in a selective way, on some of its parts, thus increasing our ability not only to act on our extraphysical manifestation, but also to create a personal energetic condition which is more favorable to our consciential evolution.

While awaiting for the author's future works, I'm certain that *Vibrational State and Energy Resonance* is a text that I will study and practice for quite some time, and I can only encourage its readers to do the same, be they scholars or lay people with some genuine interest in the research and development of their multidimensional consciousness.

Previous volumes